THE GUINNESS BOOK OF *More* MILITARY BLUNDERS

GEOFFREY REGAN

GUINNESS PUBLISHING

For Fay Russell

Published in Great Britain by
Guinness Publishing Ltd
33 London Road, Enfield, Middlesex EN2 6DJ

First published 1993

ISBN 0-85112-728-2

A catalogue record for this book is available from the British Library

Designed by Cathy Shilling

Typeset by Ace Filmsetting Ltd, Frome, Somerset

Printed and bound in Great Britain by
The Bath Press, Bath

Front cover illustration: The Battle of Waterloo, 18 June 1815
(Archiv für Kunst und Geschichte)

PICTURE ACKNOWLEDGEMENTS
The Publishers wish to thank the following for permission to reproduce pictures in this book: Archiv für Kunst und Geschichte; Associated Press; Image Select; Imperial War Museum; Keystone; Mary Evans Picture Library; National Maritime Museum; Peter Newark's Military Pictures; Popperfoto; Press Association.

CONTENTS

INTRODUCTION

In writing about military incompetence it is only fair to say at the outset that I am not suggesting for one moment that military commanders deliberately set out to fail. In the words of Ben Jonson, I intend merely to 'sport with human follies not with crimes'. Yet there is no denying that military follies are generally tragic, so much so that many would like to call them crimes. The slaughter on the Somme or at Passchendaele is sometimes viewed as a 'crime' rather than the product of inept leadership. But if crimes presuppose intention then it cannot be said that men like Haig and Gough, or indeed Napoleon, Robert E. Lee and Ulysses Grant – all men who squandered lives at some point in their career – intended their men to be slaughtered. The most we can accuse them of is 'failure to take due care and attention'. And there is far more weight in this charge. Throughout this book – and others I have written – are examples of commanders taking far too little care of their men. At the beginning of the First World War it seemed to be assumed that manpower was so abundant that it could sustain any amount of casualties, but by 1917 all the combatants in the West knew that this was not true. Only then, with manpower becoming short, and precious, did both sides begin to innovate and strive for ways of winning the war that did not involve a heavy human cost. The art of generalship was reborn. There were still mistakes made in 1918 but they were far fewer in number.

In Chapter One I examine an area of military incompetence that has never before received much attention. It is the subject of 'friendly fire' or, as the Americans now call it, 'amicide', or 'blue on blue'. This has been an ever-present problem in warfare since men first threw rocks at each other and hit their wives by mistake. But since the advent of killing at long range, through artillery, rockets, ground-support aircraft and helicopter gunships, the problem has become a very important one. In both world wars a substantial number of casualties on both sides were sustained as a result of 'friendly fire'. In many ways amicide is the most complete form of military incompetence; to kill one's comrades-in-arms rather than one's enemies must be the most basic mistake any soldier can make. As warfare has become more long-range and indirect – as well as more destructive – the chances of a commander calling down artillery fire or calling in an air strike on the wrong map coordinates must be very great. It has even been suggested that 'friendly fire' will cause an increasingly large percentage of all battle casualties in future wars. The losses in the Gulf War from 'blue on blue' strongly support this prognosis.

Chapter Two explodes the myth of Britain's skill in combined operations. Far from being a successful form of warfare for Great Britain, combined operations have produced some of the most famous military disasters in her history. Why this should be is a question that is difficult to answer. Combined operations have seemed to involve the pooling of weaknesses rather than strengths. The navy has rarely been on top form during such operations, sometimes landing troops at the wrong places – as at Santa Domingo in 1654 and more recently at Suvla Bay in 1915. Naval commanders – notably Admiral Vernon at Cartagena and the Earl of Cork at Narvik – have often failed to see eye-to-eye with their military counterparts, with the result that cooperation has been scanty. Nor has the army covered itself with glory on such occasions. Opposed landings as at Camaret Bay and Ile de Ré have frequently brought out the worst in soldiers who seemed to regard water as an element completely foreign to them, causing them to panic and flee. At Jamaica in 1654 it was the rustling sound of crabs that set their legs working in reverse. When besieging coastal forts British soldiers seem to have experienced an inordinate amount of trouble with ladders. They were either left behind or stolen or even, in two cases, cut too short so that the assault troops could not even reach the parapets. Even as late as the Suez landing in 1956 and the Falklands War of 1982, vital supplies were still stowed aboard transport vessels without any thought for when they might be needed. Should it be thought that the British alone have been incapable of launching a combined operation that did not turn its commanders grey overnight and result in total failure, one should point out that the Americans at Santiago de Cuba in 1898 and the French in Madagascar in 1895 were equally inept.

Military incompetence, as we have seen, can take many forms. Inadequate weapons and uniforms, poor food, faulty transport all play their part in condemning an operation to failure. But in the 20th century, with the advent of air warfare, a new problem has arisen: the suitability and effectiveness of aeroplanes. While German commanders have only rarely come under scrutiny for incompetence in the past, the Germans have made a significant contribution to the phenomenon of friendly fire. Germany features rather more prominently in the aeronautical chapter than one might expect such a technically competent nation to do. Even the able Professor Messerschmitt was capable of designing an aerial horror like the 'Gigant'. British, American, French and German designers were able to think up planes that lost their wings when the wind blew, planes that stopped you seeing when you came in to land and planes whose propellers cut your head off if you leaned back.

I have honoured the 'Sawdust Caesars' of the 20th century with a chapter of their own. We can see how Hitler's chauvinistic attitude to women cost him three million men on the eastern front, and how his over-excitement at winning a battle cost him the war. The original 'Sawdust Caesar' – Benito Mussolini – wanted so much to be a great military leader that he overruled his generals and invaded Greece at the very time when he was demobilizing his army to pick the harvest. In Britain the much-vilified but outstandingly imaginative Winston Churchill could not resist interfering at all the wrong moments. He was always trying to eat the cakes before they had cooled. Some strict Nanny figure – probably Alanbrooke – should have smacked his hand over Greece and told him to leave it to the Greeks. Moving on from Churchill brings us inevitably to Margaret Thatcher and the Falklands War. However much one admires her handling of the war itself one must question the handling of the crisis that made war necessary. And last in our survey of political figures who could not be prevented from interfering in military matters we come to Saddam Hussein – in truth neither a politician nor a military man. Yet Saddam set himself up as a new Saladin and used the threat of military power as a bluff to help him negotiate with the West, notably the United States. In the end he miscalculated the West's reaction and suffered a defeat from which it is doubtful if Thatcher and Churchill – even feasibly Mussolini and Hitler – would have survived. Yet Saddam, at the time of writing, has survived.

But as I have said, the book contains more fools than rogues. Who can fail to sympathize with the invading French general Humbert as he wanders round the Irish town of Killala trying to get somebody to take notice of him? Or the cowardly General Venables who hid behind a tree and preferred the company of his wife to that of his soldiers. Who can fail to commiserate with the two idiot generals at Dreux – Condé and Montmorency – each made prisoner by the other side, or with Charles the Bold of Burgundy, who was willing to go to any lengths to become a great general, but condemned by nature to be a complete incompetent? We see their like in all kinds of professions where they can do little harm to themselves or to others. It is only when such men – and others like the insane Captain Voulet, or the idle General Cadorna – are given military commands that their capacity for mischief multiplies a hundred or a thousand fold. Then it is that we condemn them for the damage they do and the suffering they cause.

'By God! If there is a mutiny in the Army – and in all probability we shall have one – you'll see that these new-fangled schoolmasters will be at the bottom of it.'

The Duke of Cambridge, 1856

Opposition among traditionalists, of whom the Duke of Cambridge was the most prominent, to anything resembling the education of the British soldier left Britain trailing behind such 'progressive' states as Prussia, Austria and France in the creation of a modern army.

THE BATTLE OF SAN JUAN HILL (1898)

The American campaign in Cuba in 1898 has been compared to a circus or a Wild West Show rather than a military operation. All kinds of specialist acts were there, from Colonel Derby with his inflatable hot-air balloon and Theodore Roosevelt's 'Rough Riders' to the world's fattest man, General William Shafter, a 30-stone monstrosity, variously described as 'a vast floating tent' or 'three men rolled into

A flamboyant depiction of Theodore Roosevelt and his 'Rough Riders' during the Spanish-American War of 1898, by magazine artist W.G. Read. Their famous charge at the battle of San Juan Hill was, in fact, on foot.

one'. Probably the best act was a decrepit old man with white hair and a beard who had apparently fought on the Confederate side in the Civil War. Known as 'Fightin' Joe' Wheeler, he kept the troops in fits with his comic impersonation of a general.

In fact, the war against Spain in 1898 found the United States totally unprepared for the demands of modern war, either on land or at sea. Fortunately for the Americans, Spain was just as bad, being militarily third-rate and even more confused and inept in its preparations. The crucial land fighting took place in Cuba, and Major General William Shafter was selected to lead an American expedition there to capture the garrison and port of Santiago. Shafter was a curious choice for such an arduous task, being elderly, gouty and – at 400 pounds – grossly overweight. In the oppressive heat of Cuba, Shafter was to become little more than a bystander, and much of the fighting seemed to take place by itself, without direction or planning.

America's preparation for the Cuban expedition was marked by scenes of unbelievable chaos. Although Tampa, Florida, the choice of port for embarkation, was close to Cuba it had few other advantages. Its one pier and single-track railway were quite inadequate for the embarkation of 17,000 men. Trains were obliged to queue up for many miles and freight cars arrived in no particular order and without any sign of what was inside them. One officer looking for bacon and beans found only shirts and shoes, while another seeking artillery parts found biscuits. There were so few ships to transport the troops that the men fought each other for places rather than miss the fun. Supplies and ordnance were loaded without any consideration for the fact that the Americans might face opposition to their landing on the beaches in Cuba. The whole business was conducted in the atmosphere of a fairground or children's party. In the end, 10,000 troops had to be left behind for lack of space.

A German military observer, head in hands, witnessed the disembarkation at Daiquiri in Cuba. He remarked that 300 determined men could have stopped the Americans from reaching the shore at all. But as was to happen frequently in the next few weeks, the Spanish missed their opportunity to punish the Americans for their amateurishness. In the absence of clear planning the troops got ashore as best they could, but no one had thought about the problem of landing the cavalry horses. Eventually the creatures were thrown into the sea and left to swim ashore, though some swam out to sea and were drowned. Even though the Americans were landing on a hostile coast they came ashore at night under navy searchlights and without any attempt at concealment. Some soldiers were laughing and shrieking with the excitement of it all, while others played in the surf as if it were Coney Island. An advance force under General Lawton moved down the coast to take the port of Siboney, where further landings took place. Again the Spanish did nothing to oppose them. The American soldiers were disappointed at the lack of fighting spirit shown by the Spanish. Their idea of tactics was revealed by one sergeant:

> There's Santiago, and the dagoes, and here we are, and the shortest distance between two points is a straight line; which is something everybody knows, and don't have to study strategy to find out. I am in favour of going up there and beating the faces of them dagoes.

Whether General Shafter needed such advice we will never know because nobody took any notice of him throughout the campaign. The 'Rough Riders', led by Colonel Leonard Wood and Republican senator Theodore Roosevelt, were a law unto themselves and seemed to do as they pleased, while ex-Confederate General Wheeler was determined to live up to his epithet of 'Fightin' Joe'. Defying orders from Shafter, Wheeler rushed off one night with 1,000 men looking for the nearest Spaniards to fight. Finding them at Las Guasimas, he immediately got into difficulties and had to send back for help. A quite unnecessary battle followed, ending in an American victory, but at a cost of 116 lives. The Americans seemed to have enjoyed themselves: Wheeler, momentarily forgetting what war he was in, yelled, 'We've got the damned Yankees on the run' as the Spaniards withdrew, and Roosevelt squeezed every ounce of political advantage from his minor part in the affair, emerging as a national hero instead of being court-martialled for disobeying orders.

Shafter now caught up with his advance guard and made preparations for an assault on Santiago. He decided on a largely unnecessary attack on the Spanish village of El Caney prior to assaulting the fortified San Juan heights. But before any of this could happen, he collapsed from combined attacks of gout, heatstroke and malaria. Delirious as he was, Shafter still believed he could direct affairs from his sickbed, and so he sent orders through his ADC, Lieutenant Miley. In theory this could have led to a disaster, as it left vital on-the-spot decisions to a junior officer unequipped to make them. In practice, however, the men did very much what they liked and so no real harm was done.

The attack on El Caney, planned to be 'bloodless' and take just two hours, was actually a 'bloodbath' that went on for eight. Although the Spaniards were outnumbered eleven to one, their fortifications were stronger than the Americans had expected. The pointless assault on El Caney tied up an American division for the best part of a day and meant that the main attack on San Juan was first delayed and later begun without them. American casualties at El Caney – 81 killed and 360 wounded – were again quite unnecessary.

The assault on San Juan Hill took place next and was characterized by errors and incidents so ridiculous that it is difficult to believe that it was undertaken by trained soldiers at all. An artillery battery under one Captain Grimes, firing without the use of smokeless powder, attracted a hail of Spanish shells on itself and the crowd of soldiers and Cuban irregulars watching the pyrotechnic display. It was soon obvious that Grimes was merely acting as a target for the Spanish gunners. He was ordered to cease firing and stomped off in a sulk. The infantry, it was decided, would have to go in without artillery support. American technical wizardry was represented by an inflatable hot-air balloon in which Lieutenant-Colonel Derby ascended to spy out Spanish positions. As the infantry advanced through the thick jungle, Derby's balloon floated in the air above them. The Spaniards were not cowed by this display of American aerial power, and merely used the balloon as a way of picking up where the American columns were at any one moment in the sea of green foliage below them. The American troops, who had presumed themselves well hidden, found themselves under heavy and accurate fire from the Spaniards on San Juan Hill, who did not need to see through the jungle canopy but just followed the balloon. Shafter, from his position a mile behind, also found Darby's balloon useful as otherwise he had not the faintest idea where his men had got to. It was as if the jungle had swallowed them up.

The march along the jungle paths was led by the 71st New York Regiment, who were resentful because they could not have the modern rifles some other regiments had got and whose old firearms used black powder that made them easy targets for the enemy. Under fire from an unseen foe they faltered and then turned to run. Eventually, their officers had to let them lie in the long grass while the others troops passed them by. When the American troops came out into an open meadow below the San Juan heights they discovered that the Spaniards had erected barbed-wire obstacles. Since no one had thought to bring wire cutters they had to climb over the wire as best they could, with some of them using bayonets to hack at the wire or the metal posts. The American infantrymen now faced an uphill assault without artillery support against entrenched opponents – a proposition at which many trained soldiers would have quailed but which seemed to the American 'dough-boys' just another part of the obstacle race.

At that moment an American battery of Gatling guns arrived on the scene, spurring the 71st New Yorkers to let out a great roar of delight. Unfortunately, this gave their position away, and the Spaniards poured a devastating fire into the greenery from which the cheering had come. Many of the 71st were put 'forever beyond the possibility of cheering'. As the other American troops fought their way through the barbed wire and up the slopes some of the Spanish defenders gave way and fled. Captain Grimes, who had been brooding over having been left out of the battle, at last saw his chance to get some of the action – to disastrous effect. Seeing tiny figures on the hills, Grimes immediately assumed them to be Spaniards and ordered his battery of guns to open fire on them. A torrent of shells now fell on the advancing Americans, forcing them to fall back down the hill. An officer who tried to stop the gunners by waving his hat at them was hit by a shell splinter for his pains. Happily for the attackers the Spaniards again missed the chance to counter-attack and rout the confused American infantry. Once Grimes was prevailed upon to stop firing, the Americans overran the Spanish defences by sheer weight of numbers. Yet in spite of a numerical superiority of sixteen to one in the attack on San Juan Hill, the Americans had done their best to fail, losing 208 killed and 1,180 wounded. As streams of injured men were led down the hill it was discovered that there were just three ambulances available to deal with them.

There can be no doubt that Shafter's physical weakness contributed to the chaotic handling of this battle, yet the American government must have known of his incapacities before he was sent to Cuba. The Americans eventually achieved victory in spite of their commander. The raw courage of many of the troops made up for the limitations of their officers and their outdated equipment. Ironically, the assault at San Juan Hill was probably unnecessary. If Shafter had employed the guns of Admiral Sampson's fleet, which could have bombarded the Spanish defences from a safe distance of 8,000 yards, the Spaniards would very probably have been forced not only to evacuate San Juan Hill but also to surrender Santiago itself. But as 'Fighting Joe' and the 'Rough Riders' would have said, why should the navy have all the fun?

'There is a time for everything, and the time for change is when you can no longer help it!'

The Duke of Cambridge, 1890

Condemned out of his own mouth, the dunderhead duke conducted a lengthy campaign to resist all attempts to modernize the British Army. For decades he opposed every effort by the progressive and scientific Sir Garnet Wolseley to adopt the latest improvements in weaponry and tactics.

Part I

CHAPTER 1: AMICIDE – THE GENTLE ART OF KILLING

The Americans have a word for it – amicide – but to everyone else it is known by that ugly euphemism 'friendly fire'. In simple terms it means killing and wounding members of your own side in the chaotic environment of the battlefield. It has always been present as an unavoidable factor of war, from the days of Alexander the Great and earlier when the rear-rank soldiers of the Greek phalanx were sometimes impaled by the sharp spear butts of the front-rank men. In some ancient battles, notably at Cannae in 216 BC and Adrianople in AD 378, soldiers were so crowded together that they could not reach their enemies with their weapons and their blows fell instead on the helpless heads of their colleagues. Failure to take account of wind conditions often resulted in arrows falling among friends as well as foes, as happened at Towton in 1461, during the Wars of the Roses.

The effects of friendly fire are often even more devastating than those of enemy action, as the troops who suffer have no expectation of attack from that direction and consequently make no attempt to protect themselves. When it occurs, friendly fire is generally the result of either weapon mishandling or malfunction, or the faulty identification of targets. In the case of the latter, human frailty – through the misreading of grid references or mishearing of orders – has often resulted in the artillery bombarding friendly positions. Even more dramatic have been the numerous examples of ground support bombers misreading maps, wrongly identifying smoke signals or even getting the date wrong. During the campaign in Normandy in 1944 disgruntled American GIs named their own aircraft the 'American Luftwaffe' because of the number of times friendly casualties were inflicted. It has been estimated that as much as 2 per cent of all American battle casualties during the Second World War were caused by 'friendly fire'. The figure was almost certainly higher in the Vietnam War, where 'Stormin' Norman' Schwarzkopf himself was almost killed by 'friendly' or 'blue on blue' bombing from B-52 bombers, and a young soldier in his company was actually killed by American artillery fire. An angry Schwarzkopf has since been at pains to condemn the expression 'friendly fire', pointing out that no bullet that leaves the barrel is ever friendly, regardless of who pulls the trigger.

The most shocking manifestation of amicide – which reached its apogee in Vietnam but has been a feature of all modern wars – is the deliberate killing of officers and NCOs known as 'fragging'. It is frequently the result of a breakdown of discipline and as such is relevant to our study of military incompetence. In Vietnam, between 1969 and 1972, it has been conservatively estimated that there were 788 cases of 'friendly assaults with explosive devices'. This was just the tip of a very big iceberg. Only a tiny proportion of these murders were ever reported.

The best of enemies

The battle of Waterloo in 1815 has always been famous for the close cooperation shown between the Duke of Wellington's forces and Marshal Blücher's Prussians. What is not generally known is that as soon as the Prussian artillery reached the battlefield they opened fire on the British horse artillery by mistake, causing more casualties than the French had. The British artillery responded in a perfectly natural way; they fired back at the Prussians so that the battle against Napoleon turned into a three-way contest. Eventually a tall German officer rode over to where Captain Mercer's horse artillery was stationed, calling at the top of his voice, 'Ah! mine Gott! – mine Gott! vot is it you doos, sare? Dat is your friends de Proosiens; an you kills dem! Ah mine Gott! – mine Gott! vill you no stop, sare? - vill you no stop? Vat for is dis? De Inglish kills dere friends de Proosiens! Vere is de Dook von Vellington?' Mercer, unimpressed by the man's accent, remarked that it had been the Prussians who fired first. He agreed to stop firing for a while until, with the German officer standing beside him, they were both forced to dive for cover by more Prussian shells. Mercer now told the German to ride back and tell his friends that the British would stop firing when they did – and not before. The bemused German returned to his horse, muttering 'Oh, dis is terreeble to see de Proosien and de Inglish kill vonanoder!' He rode off and after a few further rounds had been fired the Prussians ceased their bombardment and the British gunners were able to return to their task of fighting the French.

Lee's right arm

General Thomas 'Stonewall' Jackson was the most renowned of Robert E. Lee's lieutenants in the American Civil War. Lee claimed that it was Jackson – his 'strong right arm' – who contributed most to his greatest victories. In 1863 at the battle of Chancellorsville Jackson had run rings round the Union commander 'Fighting Joe' Hooker and had laid the foundation for Lee's most perfect victory. Yet for Lee the battle ended in tragedy when news reached him that Jackson had been shot – by his own men.

Riding at dusk near the town of Chancellorsville, Jackson and his staff surprised some men of a North Carolina brigade who, not suspecting horsemen to be approaching from the direction of the Union army, opened fire, killing two of Jackson's companions and hitting the general with three bullets, one of which shattered his left arm. Jackson was carried by litter to a field hospital where the arm was amputated. When Lee heard the disastrous news he sent Jackson a letter congratulating him on the victory which he claimed was due to Jackson's 'skill and energy'. But morale throughout the Confederate army slumped at the news of Jackson's wounding. Worse was to follow. Pneumonia set in and three days later the great 'Stonewall' Jackson died. The death of Jackson was a blow to the whole

The Confederate general Thomas 'Stonewall' Jackson earned his nickname at the first battle of Bull Run in 1861, when his troops held off repeated Union assaults to lay the foundations for a Confederate victory. Jackson's death at the hands of his own troops at Chancellorsville in May 1863 was a blow from which the Confederacy never recovered.

Confederacy. Had he been at Lee's side at the battle of Gettysburg, a Confederate victory would have been probable on the first or second day of fighting. But for Jackson's untimely end the whole history of the United States might have taken a different course. Nor was Jackson the only Confederate general to suffer at the hands of the trigger-happy 'Rebs'. At the battle of the Wilderness on 6 May 1864 'Old Pete' Longstreet – another of Lee's most able lieutenants – was severely wounded by friendly fire.

'I cannot help wondering why none of us realized what the most modern rifle, the machine gun, motor traction, the aeroplane and wireless telegraphy would bring about. It seems so simple when judged by the results.'

Field Marshal Sir John French in 1919

This embarrassing statement was made by the man thought fit to lead the BEF in France in 1914. It reveals a mind of startling naïveté.

The first to go

Hard though it is to believe, in the first days of August 1914 casualties in the British Expeditionary Force could still be regarded as individual tragedies. The men lost still had names and identities, and their deaths were tragic and yet significant events. But such feelings took an early battering when it was revealed that the first British soldiers of the BEF to die in France had been killed by their own side.

As the 80th Battery of horse artillery, attached to the 5th Division, rode up towards the village of Le Cateau at nightfall, they sent out riders to check the road ahead. The local Civil Guard – composed mainly of untrained peasants – had been watching the road. When they saw the British riders approaching and heard them speaking in a foreign language they panicked and opened fire, spraying bullets in all directions. The British horsemen turned back and rode hard for the British lines. Unfortunately British infantry picquets had seen them coming at the gallop and, presuming them to be hostile, opened fire down the road, killing one man outright and mortally wounding another. It was incomprehensible: two men dead and 5th Division had not even seen a German yet.

Too little, too soon

The appearance of the first tanks on the Somme battlefield in September 1916 signalled one of the most important advances in warfare since the discovery of gunpowder. Yet in his hurry to use them General Haig was willing to reveal this secret weapon before it was ready. Winston Churchill, for one, believed the tank should have been kept secret until Britain had enough ready to make a decisive breakthrough in the German lines. But Haig was adamant, and as usual it was the common soldier who suffered.

The first tanks were painfully slow, vulnerable to artillery fire, liable to bog down in the heavy mud and sometimes unable to differentiate friend from foe. On 15 September there was great excitement at 4th Army Headquarters when it was reported that two tanks had broken through the German lines and reached the village of Flers. General Haig even passed the news on to the waiting pressmen. What was not reported was that one tank had wiped out all the troops in a nearby British assembly trench. The 9th Norfolks had been preparing to go over the top when the tank lumbered up, lost its bearings and confused their trench with the German front line. Its machine gun raked the trench, killing dozens of helpless men. An infantry captain ran towards the tank, waving his arms and trying to make himself understood. Peering through a slit in the armoured side the machine gunner at last understood his ghastly error; the tank stopped firing and swung away, having wrecked the British attack and left the German front line unharmed. When the remnants of the Norfolks attacked later that day they were cut down by the same German defenders who had been the tank's real target.

When used for the first time on the Somme in September 1916, the tank had a dramatic effect, creating panic in the German lines; but it was mechanically unreliable and proved vulnerable to any weapon more powerful than a machine gun.

Sir John French: *'The British Army will give battle on the line of the Condé Canal.'*
Sir Horace Smith-Dorrien: *'Do you mean to take the offensive or stand on the defensive?'*
Sir John French: *'Don't ask questions, do as you're told.'*

A recorded conversation between the commander of the BEF and the commander of the British II Corps, September 1914

Sir John French allowed his personal dislike of Sir Horace Smith-Dorrien to affect his thinking at this crucial point in the war. With orders like the above it is hardly surprising that French inspired no confidence in his subordinates or his allies.

The tragedy of Samogneux

Ground forces have always been in danger from heavy artillery firing in support of them from miles behind the lines, and this threat was never greater than during the First World War. Although comparable figures are not available for all the combatant nations, the French have estimated that more than 75,000 of their soldiers died during the four years of fighting as a result of 'friendly fire' from their own artillery. And for France the tragedy at the village of Samogneux has come to symbolize the true horror of 'friendly fire'.

On 21 February 1916 the Germans began their great attack on the French fortress of Verdun. At Samogneux elements of the French 72nd Division under Lieutenant-Colonel Bernard heroically held up the German advance for two days. Hemmed in on all sides by Germans, Bernard found it increasingly difficult to communicate with his headquarters. As he said in the last message he managed to send, 'All the horses have been killed, bicycles smashed, runners wounded or scattered along the routes. I shall be doing the impossible if I keep you informed of events.' After this, headquarters received no further word from Bernard. The only evidence of the fate of Samogneux was received by Major Becker who heard a courier riding past him at full gallop shouting, 'The Boche is at Samogneux.' Becker was unable to question the man further and so assumed that the worst had happened and that Bernard's resistance was ended. The Germans must have taken Samogneux. When this news was passed to General Herr in Verdun he ordered the village to be recaptured immediately. Before an assault could be launched, however, it would be necessary for the French artillery – 155-mm guns – to saturate the area and destroy the advanced German positions. On the night of 23 February a massive artillery barrage centred on Samogneux. Unfortunately, at the very moment that the first French shells began to land, Bernard had at last found a way of sending a message to tell HQ that he was still holding on. It was too late. Even though Bernard's men fired green 'ceasefire' rockets to try to stop their gunners it was to no avail. For once the French gunners showed unerring accuracy and the French defenders were massacred. Unknown to the gunners, they had made matters easy for the Germans who were able to walk into the village unharmed. The sight that met their eyes was astonishing. The entire garrison had been wiped out by their own artillery – that is, except one man. As they stepped over the rubble the Germans heard a small voice saying, *'Pour mes enfants, sauvez-moi!'* By an incredible chance Colonel Bernard had survived the inferno. He was rescued and brought before the Kaiser himself. When questioned by the German emperor he defiantly replied, 'You will never enter Verdun.' And he was right. Yet the defence of Samogneux had destroyed the 72nd Division, which suffered 80 per cent casualties. How many men died in the ruins of the village from General Herr's guns will never be known. But the effect of 'friendly fire' had not only cost France hundreds of lives; it also presented to the Germans a vital strategic position in the defence of Verdun.

A few of the first

The 'Phoney War' at the beginning of the Second World War gave Britain the time she needed to organize her aerial defences. Not surprisingly there were hiccups, at least one of which had tragic consequences. On 6 September 1939, a newly installed radar unit malfunctioned and indicated that a massive German bomber force was approaching. Fighters were scrambled and a tragic dog-fight ensued in which two Hurricanes were shot down by Spitfires, and one pilot was killed. To make matters worse the anti-aircraft batteries then joined in, shooting down one of the Spitfires. What might have been the result if this shambles had occurred a year later, during one of the most crucial moments of the battle of Britain, is too painful to imagine.

'One a day in Tampa Bay'

As well as fighting the Allies in Russia, Italy, North Africa and Normandy, between 1941 and 1944 the German *Luftwaffe* fought an even more terrible enemy at home – itself. It has been estimated that 45 per cent of German aircraft destroyed in this period were lost for non-combat reasons. The accident rate in planes and aircrew or those lost to 'friendly fire' reached proportions that threatened Germany's capacity to continue the war. But figures for the United States Air Force were even worse. In a single year, 1943, 2,264 pilots and 3,339 aircrew were killed in 20,389 aircraft accidents in the United States itself. The following year things were hardly better: there were 16,128 accidents, 1,936 pilots and 3,037 aircrew killed. So often were B-26s destroyed by their own crews that a popular saying at one Florida air base went, 'One a day in Tampa Bay'.

The fearsome Ju87 Stuka dive-bomber was the scourge of Allied ground troops in the early years of the Second World War. However, German troops learnt what it was like to be on the receiving end of an attack by Stukas in two 'friendly-fire' incidents near Amiens in May 1940.

Unfriendly fire

In the early months of the Second World War battlefield air support was an important factor in Germany's successful *blitzkrieg* campaigns in Poland and France. However, the close cooperation between ground and air forces was always a finely judged line and things could and did go wrong, as General von Mellenthin experienced in Poland. He remembered one occasion where a low-flying aircraft passed over his headquarters, whereupon his flak gunners – without attempting to identify the intruder – opened fire with all their guns. An air liaison officer rushed about trying to stop the gunners, telling them it was a German plane – an old Stork – but the gunners were so excited that they took no notice. By an incredible stroke of luck the old plane was not hit and landed unharmed. As the gunners – now chastened by seeing the plane's identification marks – gathered round to see how the pilot was, out stepped the Luftwaffe general who was responsible for close-air support. As von Mellenthin observed, 'he failed to appreciate the joke'.

The Ju87 Stuka dive-bomber was one of the most feared weapons of the German *blitzkrieg* in Poland and France. One German commander was to find out what it was like to experience a Stuka attack from the point of view of the victim. On 14 May 1940 a group of Stukas mistakenly attacked 2nd Panzer Brigade at Querrieu, near

Amiens. Their mistake was quickly brought home to them by the brigade's commander, General Heinz Guderian, who ordered his flak gunners to open fire on the German planes. Only days before Stukas had seriously damaged a column of German tanks near Cheméry, narrowly missing Field Marshal von Rundstedt, commander of Army Group 'A', and Guderian was taking no chances. As he said, 'It was perhaps an unfriendly action on our part, but our flak opened fire and brought down one of the careless machines.' The two crewmen escaped by parachute and landed in the midst of the German infantry who not surprisingly told the airmen what they thought of them. Guderian called off his enraged men and having himself torn the flyers off a strip, 'fortified the two young men with a glass of champagne'. Guderian could afford to be magnanimous. The Stukas had done no harm on this occasion and their work elsewhere was making things easy for the German Panzers.

An unwelcoming committee

During the Allied invasion of Sicily in July 1943 American paratroopers were dropped behind German forward positions. But the men of the 504th Regiment, commanded by Reuben Tucker, soon found that they had more to worry about than fighting Nazis. Although the Americans had gone to some lengths to ensure that their gunners could identify their own planes when they came over, everything possible went wrong. The 2,000 men of the 504th arrived over American lines just after a German air raid and the ground gunners were feeling edgy. As the 144 C-47 transports passed above them in the darkness a gunner on a nearby ship panicked and opened fire. That was the signal for an outburst of mass hysteria. Every gun, both on land and at sea, began blazing away at the American planes. Soon 33 of the transports were spiralling down in flames, while 37 others were badly damaged. In a matter of minutes Tucker's troops suffered 318 casualties, with 88 men dead. The sky was filled with parachutes, many in flames, as the crack airborne troops baled out of their stricken transports. So furious at this fiasco were the American commanders that they came close to cancelling all future airborne operations.

'George, let me give you some advice. If you get an order from Alexander that you don't like, why just ignore it. That's what I do.'

General Montgomery to General Patton, 1943

In the Italian theatre of operations in 1943 the Allied Commander-in-Chief, General Alexander, found at least two of his senior commanders – Bernard Montgomery and George Patton – highly insubordinate. Each seemed more content with running his own private war, not always to the advantage of the Allied cause.

The 'Cobra' strikes

During the Normandy campaign of 1944 a macabre joke was common among the Allied troops. It went something like this. When the Germans carry out a bombing raid the British duck and when the British carry out a bombing raid the Germans duck, but when the Americans start bombing everybody ducks. Like many jokes in wartime it contained a bitter truth. The accuracy of American ground support bombing throughout the war left a lot to be desired and casualties from such 'friendly fire' reached unforgivable proportions.

In July 1944 the American breakout from their Normandy bridgehead – codenamed Operation 'Cobra' – began with preliminary air bombardment of the forward German positions. In order to reduce the chance of inflicting casualties on his own men General Omar Bradley came up with what he thought would be a foolproof system. The bombers were given a clearly identifiable bombing line – the St.Lô to Péries road – beyond which they must not drop their bombs. They were instructed to fly parallel to but south of the road; this meant that they would not actually fly over American troops at all. It all sounded simple. But in war even the simple things can and do go wrong. For a start, the air chiefs were not going to be dictated to by a footslogging general. They rejected his parallel approach, arguing that it would take too long for all the planes to get in position in time to bomb the targets. Instead, the pilots were ordered to cross the road at right angles – just about as big a change from Bradley's plans as can be imagined. To make matters worse, nobody bothered to tell Bradley that the change had been made.

On 24 July, with the weather improved and visibility generally good, the bombers set off. Unfortunately, while they were on their way the clouds closed in again and the mission was aborted. But the news did not reach all the planes in time and some of them simply dropped their bombs through the clouds, missing their targets and killing 40 or 50 American soldiers in the process. When Bradley was informed of the disaster he was furious with what he called the 'duplicity' of the air planners, who had ordered their pilots to fly in at right angles rather than parallel to the road.

This first day's disaster should have warned the American commanders to pull back their troops from the bombing zone, but when they tried to do this the Germans promptly advanced and occupied their positions so that when the bombers dropped their payloads they found that they fell behind the Germans rather than on top of them. The real problem, as the airmen tried to tell Bradley, was that the heavy bombers were quite unsuited to ground support. They could not guarantee accuracy from 8,000 feet. Also, it must be said, the ground troops showed a touching – if misplaced – confidence in the work of their airmen and stood up waving as the bombers flew overhead. Many GIs apparently died waving their helmets at the bombs that killed them.

The following day – 25 July – the full-scale bombing raid by B-24s took place and Bradley's worst fears were now realized. In one incident the bombardier of the leading plane failed to set his bombsight properly. His faulty example was then copied by the following eleven planes so that all twelve dropped their bombs into the

American general Omar Bradley took pains to avoid casualties from 'friendly fire' during the Normandy campaign of 1944. But the incompetence of American air force commanders led to US ground troops suffering unprecedented losses at the hands of their own airmen.

American lines – a total of 47,000 pounds of high explosive bombs. Matters got worse when the next flight in failed to identify the target properly and just followed the example of the previous flight, dropping 352 260-lb fragmentation bombs on the suffering GIs below. Even this was not the end. Another pilot overruled the correct targeting of his bombardier and ordered the man to tip his bombs into the inferno that had been the American front line. The result was that 136 Americans died and over 621 were wounded. Among the dead was the commander of First Army Group, Lieutenant-General L.J. McNair, who was killed by a direct hit on his foxhole.

The whole débâcle of Operation 'Cobra' was a savage lesson for General Bradley on the limitations of Allied tactical bombing. From the top – General Doolittle in charge of Eighth Air Force – to the lowest bombardier it had been a thoroughly unprofessional operation and a tragic waste of American lives.

'I want you to get to Messina just as fast as you can. I don't want you to waste time on these maneuvers, even if you've got to spend men to do it. I want to beat Monty into Messina.'

George Patton to Omar Bradley, August 1943

Patton and Montgomery were both prima donnas whose personal rivalry often took precedence over the Allied war effort. In the above George Patton shows that he was quite prepared to squander American lives to win a pointless propaganda victory over a man – Montgomery – whom he seemed to hate even more than the Germans he was supposed to be fighting.

Operation 'Tractable'

In August 1944 the Canadians launched an armoured offensive against German troops at Caen. As in 'Cobra' the operation was to be prefaced by an aerial bombardment of German positions. On 7/8 August 1,000 bombers from RAF Bomber Command were due to saturate the forward German positions, but such was the build-up of smoke and dust that at least a third of the planes did not drop their bombs for fear of hitting friendly troops. The next day the US Eighth Air Force took over ground support and reproduced their form in 'Cobra', inflicting 300 casualties on the Canadians and Poles. Then, as if the RAF was determined to prove that anything the Americans could do wrong the RAF could do equally badly, Bomber Command re-entered the action on 14 August. At first all went well until 77 Lancaster and Halifax bombers of the second wave flew over the British lines, dropping their bombs on 'friendly' troops, causing 65 deaths and wounding over 400 men. By an incredible oversight nobody had informed Bomber Command that ground troops were identifying their positions by yellow smoke: the fact was that yellow smoke was Bomber Command's target-identification colour. The more desperately the ground troops burned their yellow flares the more the bombers rained death on them. Worse casualties were only avoided when an Auster trainer took off and flew in front of the bombers, waggling its wings and trying to draw them away from the British lines.

The 'American Luftwaffe'

During the Tunisian campaign in 1943 – at the battle of the Kasserine Pass (see p.172) – a flight of American B-17s got lost. Instead of bombing the pass they destroyed an Arab village with their bombs, even though it was more than 100 miles from the battle area, probably a record for incompetent navigation.

In Sicily the American air force inflicted casualties on friendly forces at an alarming rate. The North American A-36 Invader was a first-rate dive-bomber, but it had a depressing record of hitting 'friendly' troops by mistake. Even when the ground forces displayed luminescent identification this did not seem to stop the A-36 once it had started its dive. General Omar Bradley narrowly missed falling victim to an A-36, which dive-bombed and strafed him while he was visiting General Allen's HQ. Bradley observed that he was getting used to it; it was his third strafing that day by American planes. The A-36s were also responsible for the loss of Monte Cipolla to the Germans. A group of American GIs were desperately holding a position on the mountain until seven A-36s bombed them, killing or wounding nineteen men, destroying their last four howitzers and driving them headlong down the slopes. In the same area A-36s also dive-bombed British XXX Corps Headquarters, mistaking it for a German-held position. Eventually, when A-36s shot up an American tank column in spite of yellow recognition signals, the GIs lost patience and shot one of them out of the sky. The pilot parachuted down and was furious when he found out that he had been shot down by friendly troops. 'Why you silly sonuvabitch,' said the tank commander, 'didn't you see our yellow recognition signal?' 'Oh,' said the pilot 'Is that what it was?'

While the infantry slogged through Italy the American airmen continued to exact a heavy toll on 'friendly' troops. Heavy bomber strikes caused widespread havoc, inflicting heavy 'friendly' casualties at Venafro, during the bombing of Monte Cassino. Venafro was fifteen miles from the target area and the American planes managed to destroy the British Eighth Army commander's caravan (General Leese was fortunately not in it) and a Moroccan military hospital, wounding 40 civilians, as well as gunners from the 4th Indian Division. During the advance on Rome, American Mustang fighter-bombers strafed columns of American troops by mistake, inflicting hundreds of casualties. In October 1944 the Ninth US Air Force attacked pillboxes in the West Wall near Aachen. Not one plane reached the target and one even attacked the wrong country, destroying the Belgian mining village of Genk and causing about a hundred casualties. The US 30th Division had quite a score to settle with their air support crews. In six months after the Normandy landings they were attacked no less than thirteen times by American planes. During the Battle of the Bulge they were based in Malmédy, in Belgium, but this did not save them from being bombed by the Ninth Air Force, who killed and wounded hundreds of soldiers and civilians over a three-day period. It comes as no surprise to learn that many GIs opened fire on their own planes on sight as their only form of protection.

A Scottish lament

Another war with the Americans as allies brought another martyrdom for British troops. This time it was in Korea in 1950. On 23 September troops from the Argyll and Sutherland Highlanders captured Hill 282 from the Communist North Koreans and called in air support. They identified their own positions with recognition panels visible from the air but when the American Mustangs arrived they disregarded the signals and plastered the British position with napalm. Seventeen members of the Argylls were killed and 76 wounded, and they were forced to evacuate the hill, which was reoccupied by the Communists. Major Kenneth Muir and his 30 surviving men then fought their way back up the hill and retook it, Muir falling in the moment of victory. For his courage and leadership Muir was awarded a posthumous Victoria Cross.

'Crazy house'

During the American invasion of Grenada in October 1983, air strikes by Corsair IIs were called in against Cuban and Grenadian People's Republican Army positions east of the town of St. George's. Unfortunately the fighter-bombers attacked Fort Matthew, an old British colonial fortress that was being used as a mental asylum. As bombs and rockets exploded in and around the fort dozens of pathetic inmates ran screaming into the streets of the town. Although the fort was clearly marked as a hospital on the American maps the pilots had confused it with other hill fortifications nearby. The fact that there was a Grenadian flag flying above the building appeared to clinch the misidentification and the Corsairs roared into action, killing 21 helpless people and wounding hundreds of others. It was a shocking and thoroughly discreditable incident.

Not content with their assault on what the Americans dubbed 'Crazy house' the Corsair pilots now turned their attention to their own ground troops. Second Brigade's Tactical Operations Centre (TOC), which was coordinating the ground operation, suddenly came under attack from more Corsairs. Marine target spotters had apparently come under fire from somewhere near 2nd Brigade's headquarters

American troops fire a 105-mm howitzer against small-arms fire near Point Salines, Grenada, October 1983. The US invasion of the tiny Caribbean island witnessed several instances of 'friendly fire', most of them the result of faulty target identification and poor mapwork.

and through mistaken target identification they had called in an airstrike on it. Luckily there were no deaths, although sixteen men were seriously wounded, one radio operator losing his legs. To prevent 'friendly' casualties all combat units had been told to show coloured smoke, but nobody had told the headquarters staff that they were a combat unit and needed to show smoke. So TOC showed no smoke and paid a heavy penalty. The Corsair pilots – roaring in at 400 mph and at treetop level – saw no smoke and fired on instinct. It was all over in a second – yet more proof, if it was needed, that the more impersonal the weapons men use the more likely it is that mistakes will be made.

Blue-on-blue

On 26 February 1991 the British forces in the Gulf suffered their heaviest losses of the entire war and, as fate would have it, they were victims of 'friendly fire' inflicted on them by American pilots. There are conflicting accounts of how this 'accident of war' occurred; without attempting to allocate blame, it is clear that this was yet another in a long history of tragic losses inflicted by carelessness in weapons handling or in identifying targets.

A company of the Royal Regiment of Fusiliers travelling in Warrior armoured vehicles in the Iraqi desert was attacked by American A-10 tankbusters, which fired maverick missiles at them, killing nine men and wounding seven. Apparently the

Warriors had been correctly marked and were within their allocated area of operation. But this did not prevent the accident occurring. The Americans were vehement in refuting charges of irresponsibility or negligence. They claimed that their planes were wrongly directed by a British air controller and that the identification marks on the Warriors had been obscured by dust. The British, on the other hand, claimed that the marks on the vehicles had not been obscured and that from an even greater height an American reconnaissance plane had had no difficulty in identifying the Warriors as allies after the attack. In such a tragic situation there can be no winners. Whatever the claims of the greatest of generals, war is littered with human errors – both large and small. As weapons become more complex and more deadly, mistakes of the kind described in this section of the book are bound to form an increasingly large proportion of casualties.

This was not the only 'friendly-fire' incident involving British and American forces. On 27 February – the last day of the war – two British Scorpion reconnaissance vehicles, crewed by men from the Queen's Royal Irish Hussars – found themselves suddenly under fire from American tanks. Two men were wounded, and once the firing ceased it was found that the Americans had scored a bullseye on the Hussars' fluorescent identification sheet.

All that we are left with is a faint uneasiness about the number of times that the 'culprits' in 'friendly-fire' incidents have been American. As recently as the Gulf War 25 per cent of American battle casualties – some 35 deaths – were caused by members of their own armed forces. It has added a worrying dimension to military training.

A British armoured vehicle of the type that was mistakenly attacked by US aircraft during the 1991 Gulf War. Nine British servicemen lost their lives in a particularly tragic example of 'friendly fire'.

Fragging

Mutiny in the British or American armies has been a very rare event in modern times. Yet mutiny can take many forms and one that has become more widespread in the 20th century is known as 'fragging', that is the murder of officers or NCOs who show an over-eagerness for seeking contact with the enemy or undertaking other dangerous missions. During the First World War, young officers – often no more than eighteen-year-old lieutenants, fresh from public school – had to order their men to go 'over the top' in the face of a withering enemy fire. It was relatively simple for the officer to be shot in the heat of battle so that his men could remain in their trenches for lack of anyone to lead them. In the Vietnam War it has been estimated that as many as 20 per cent of officer fatalities were the result of fragging, either through shooting or more often by an apparently misdirected grenade.

The murder of unpopular officers, often in action, has been widely documented even as far back as Roman times. One famous example from the 18th century involved a major of the 15th Foot Regiment at the battle of Blenheim in 1704. The man was an infamous martinet who had treated his men harshly in the past and was thoroughly hated. Aware of this he addressed his men before the battle, saying that if he should fall in the coming fight at least let it be by an enemy bullet. A soldier replied that they had more to think about than him at that moment and so the regiment went into battle. After victory had been won he turned to his men, raised his hat and called for a cheer, only to fall with a bullet through his head. During the First World War one notably unpopular sergeant was despatched when one of his men came up behind him and dropped an unpinned grenade down his trousers, blowing him to pieces.

'It is to me like a shooting expedition with just a spice of danger thrown in to make it really interesting.'

Captain Edward Hutton of 60th Rifles, commenting on the Zulu War, 1879

Both the French and German High Commands condemned the British for their amateurism and their treatment of war as merely another form of international sport. This quote – by no means an unusual observation by a British officer of the time – suggests that they were not far from the mark.

CHAPTER 2: AMPHIBIOUS OPERATIONS

The problem of divided command has been at the root of most disastrous amphibious operations. With such a long history of naval power it might be supposed that Britain had developed some expertise in this vital aspect of warfare. Unfortunately, successful combined operations have eluded Britain almost as often as they have other less noted maritime powers, and the record of British admirals and generals over the last four hundred years has been quite deplorable. Instead, Britain has performed better in the field of maritime evacuation – a notable skill as demonstrated at Gallipoli, at Dunkirk and on Crete – but, as Winston Churchill once commented, not one likely to win any wars.

Nevertheless, this has not stopped Britain continuing to plan and operate amphibious operations – often quite ineptly – even as recently as at Suez in 1956 and the Falklands in 1982. Left to their own devices the army and navy seem to be able to operate their affairs quite well, but put the two together and sailors lose even their most basic nautical skills while soldiers behave as if they have been torn from their natural element. During the Suvla Bay landings in August 1915 previously capable naval officers behaved like boys sailing their first toy boats on the Serpentine. Destroyers towing lighters full of troops anchored a mile from where they were supposed to and landed their men on the wrong beach. Others were mistakenly landed at the base of a cliff. Landing craft ran aground on shoals, while no water whatsoever was landed for two whole divisions who were without water for 24 hours. When a water lighter was sent in to save the men from dying of thirst it ran aground on a shoal 100 tantalizing yards from thousands of parched and desperate men. Troops who had driven the Turks back from their trenches were forced to wander back to the beaches looking for a drink. Combined operations have been like that sometimes. Still, one thing the navy could be relied on to do and that was to evacuate you safety when the whole operation had fallen to pieces.

'Hang the guides or hang the General'

Combined operations stand or fall on the harmony between army and navy commanders. Never was discord more strident than during the famous British expedition to Cartagena, a port in what is now north-western Columbia, in 1740–1. Relations between Vice-Admiral Edward Vernon and Brigadier-General Thomas

Wentworth plumbed such depths that the military encounters with the Spaniards were as nothing compared to the acrimony between the two men. Vernon was never known as a harmonious colleague – in fact, his nickname was 'the angry admiral' – while Wentworth suffered from the deeply singed pride of one who had been through the fires of Vernon's cooperation and emerged smouldering at the edges. At the outset, Wentworth had only been second-in-command to Lord Cathcart, who had died on the journey out. He was not himself an ambitious man and preferred the certainties of the drill ground where he could shine to the pressures of overall command on active service where he floundered. To compound his difficulties he held Vernon in a kind of awe and deferred to him on every occasion until, too late, he realized his error. He even allowed the admiral to borrow his soldiers to make up for shortages in naval manpower. Vernon clearly thought he could command the inexperienced Wentworth and bend him to his will. But Wentworth, though no match for the impatient and intolerant admiral, had a kind of stubbornness born of a close adherence to military rules and regulations. He tried to conduct his land operations entirely by the book, which infuriated Vernon to the point where he decided he really should be in charge of both sailors and soldiers. Thus instead of sharing joint command of a combined operation Vernon behaved as commander-in-chief and acted as if the army was there merely to serve the needs of the fleet.

Aware of the dangers of sickness that accompanied the onset of the rainy season in the Caribbean, Vernon was all in favour of a quick landing and an assault on Cartagena, but Wentworth did not agree. When he consulted the manual he found that it was necessary for troops, once landed, to dig in and prepare for a counter-attack. Thus Vernon and his nautical cronies looked on as Tierra Bomba, a sparsely garrisoned island guarding the entrance to the harbour of Cartagena, was fortified. For two days all was activity. Men burrowed away, sending the earth flying out of holes in the ground as if a whole regiment of dogs was hunting for lost bones at the same time. All the while Wentworth's camp was ringed round by double and triple sentries as if he expected to be assaulted from the rear. The few Spaniards on the island, feet up and reflecting on the futility of Anglo-Saxon military manuals, occasionally let off a few musket shots to remind everyone that they were still there. It was all too much for Vernon, who wrote a peremptory letter to Wentworth of a kind that can rarely – if ever – have been sent by one commander to another. Addressing Wentworth as if he were a particularly obtuse infant, Vernon suggested that he should 'push forward part of [his] force against Fort San Luis', that he should 'put the rest of [his] men under canvas', further that he should 'hasten [his] engineers to the siege of the fort', and that he should use just a few men as sentries instead of his whole army. Wentworth, miraculously, did not explode on receiving what must have read like an extract from a child's primer in military tactics. The problem was that he knew something that Vernon did not. The army had set out with just one specialist engineer officer competent to conduct a siege – Jonas Moore – and he had not yet arrived at Cartagena. The other so-called engineers, as Wentworth knew only too well, could not be trusted to build a sandcastle. Nor were his artillerymen real gunners. They had already set up their battery in direct line with the British camp, so that every time the Spaniards fired, the shots either hit them or fell amongst the soldiers' tents, killing and wounding a hundred men on the first day. Wentworth was too ashamed to admit all this to Vernon.

Jonas Moore's arrival a few days later brought further bad news for Vernon. Surveying the feeble Fort San Luis, Moore announced that he would require 1,600 men to construct a siege battery and that he would need to cut down a small forest to provide the wood. Wentworth demanded his soldiers back from Vernon, who replied that they could not be spared. A tug-of-war now began with several regiments of British redcoats as the prize. Moore, meanwhile, got on with the job of siting his twenty 24-pounder cannon. While his superiors squabbled he got the job done and the bombardment began. However, no sooner had the British guns opened fire than the Spaniards replied, killing Jonas Moore with their first shots. Vernon reacted angrily, berating Wentworth for failing to take such 'a paltry fort', and pointing out the imminence of the rainy season, which would put a stop to operations. When Wentworth tried to tell Vernon that the guns had not yet breached the walls of the fort, the Admiral insisted that the enemy would run if the fort were attacked. In fact, Vernon was right. No sooner did the British redcoats approach the walls than the Spaniards took to their heels without firing a shot. Soon the Union flag was flying over the fort, but at what cost: only 130 battle casualties but 850 either dead or sick from disease.

Wentworth's troops were quickly re-embarked for the short journey round the harbour. Just one remaining fort – Fort St. Lazar – stood between them and their goal, the city of Cartagena. Wentworth now asked for 5,000 men to be landed for the attack on St. Lazar, but Vernon told him that 1,500 was all he needed and that disembarking more would simply waste time. Hardly reassured, Wentworth advanced towards the fort, brushing aside light Spanish opposition. So feeble was Spanish resistance at this stage that an immediate attack on the fort followed by an advance on the city would almost certainly have given Wentworth both prizes at once. But the cautious Wentworth was suspicious of the ease of his progress; he expected to face further ruses and ambushes. Instead of seizing what the admirals aboard their ships saw as a golden opportunity, Wentworth set up camp several miles from the fort and prepared for a siege of Trojan duration. Vernon wrote to him: 'Delay is your worst enemy . . . We hope that you will be master of St. Lazar tomorrow.' Not to be hurried, Wentworth summoned his officers to a council of war which concluded that the fort could not be taken by assault, for its walls were too high. It would have to be besieged. He therefore asked Vernon if the fleet could carry out a bombardment for him. Vernon, believing that his ships had carried the brunt of the fighting so far and that it was now the army's turn, replied by condemning Wentworth for not having attacked the day before. Shaken by the admiral's letter, Wentworth now made a complete U-turn, and decided on an assault against the unbreached fort. However, through an administrative blunder, no tents or tools had been landed, and sleeping in the open, without cover, more and more of his men were succumbing to illness. With his only qualified engineer dead and without tools from the fleet, he was unable to erect a siege battery. There was, Wentworth decided, no alternative but to make a night assault, attacking the north and south walls of the fort simultaneously. As often happens with night attacks, particularly if the commander is weak and the troops inexperienced, everything went wrong.

The main column of some 1,000 men under Colonel Wynyard was guided towards the weak southern wall of Fort St. Lazar – where there was no ditch – by two Spanish deserters. Meanwhile, another column of British troops commanded by Colonel Grant, and accompanied, curiously, by Wentworth himself (though not apparently in a command capacity), were to make a feint attack on the hornwork to the north of the fort. Both columns began their advance at 3.00 am and Wynyard's men immediately ran into trouble. In the darkness they missed their way and found the ground far steeper than they had been led to expect, so much so that they were forced to go forward on hands and knees. They were in fact clambering up the steep eastern slope of the hill, which was protected by three lines of trenches. The British grenadiers led the assault but, as they called for scaling ladders to mount the walls, it was found that the men who were carrying them had been placed at the rear of the column. When a search was made for these soldiers it was soon discovered that they had taken to their heels, and – what was worse – had taken the ladders with them. The grenadiers, meanwhile, were just 30 yards from the walls when suddenly the Spaniards opened fire on them from point-blank range. There was no option now but to rush the walls, regardless of casualties, but who was there to lead them? Wynyard seems to have been paralysed. He and his officers, quite unaccustomed to being in action, behaved as if they were on the parade ground, advancing slowly in perfect order and attempting to fire volleys. In the murderous hail of fire the redcoats simply stood, reloaded and fired as they had been taught to do on the practice range. No officer showed any initiative and the men were bowled over in droves as the Spaniards poured grapeshot and musket fire into their thinning ranks. The grenadiers, showing admirable coolness under fire, threw their grenades into the fort, but fewer than one in three of them actually exploded – through faulty design their casings were too thick. There was no alternative but for Brigadier-General Guise, who was supervising the whole operation, to recall Wynyard's column.

Colonel Grant's column on the north face fared little better, Grant himself being shot down in the first few seconds. To fire their volleys the British infantry lined up like skittles in an alley and the Spanish cannon cut swathes through them. The call for British artillery support fell on deaf ears as the cannon had been left far behind the assaulting columns. For over an hour, on both sides of the fort, the British redcoats obeyed orders, laying down their lives in a hopeless cause. Without leadership the men went through their drills like clockwork soldiers. As day broke, the guns of Cartagena joined the fight and poured shot into British lines that were now hopelessly exposed. But still the redcoats refused to give way, standing their ground until Spanish infantry issued from the city to try to cut off the remnants of Grant's command. Only now did Wentworth order his men to withdraw, having suffered nearly 50 per cent casualties, with the officers providing tempting targets for Spanish sharpshooters. The mortally injured Grant's last words were recorded as, 'The General ought to hang the guides and the King ought to hang the General.'

Wentworth had mismanaged the whole affair. And yet responsibility for the greatest blunder – that of launching the attack without first securing a breach – lay with Admiral Vernon. Wentworth had at least called a council of war. Apart from battle casualties, which were bad enough, yellow fever was ravaging the British force. By 10 April their effective strength had shrunk to fewer than 3,200 men. The best of the old line regiments had been killed or wounded in the fighting and the survivors could hardly be trusted with any duty at all. Wentworth told Vernon that unless he agreed to lend him sailors and engineers to construct a battery to attack Fort St. Lazar he saw no hope of taking it. But Vernon was giving nothing away. If the expedition was going to fail there must be no doubt where the fault lay. The navy had done its job. If the army did not know how to besiege such a paltry fort then it was hardly up to the sailors to show them. Indomitably unreasonable as always, Vernon would take what was left of the expedition back to England and let the government decide who had failed in his duty. By the time the fleet sailed away from Cartagena there were just 1,700 men fit for action out of the original force of 9,000, and of the men who had sailed from England under Cathcart and Wentworth over 90 per cent had died on the expedition.

'It's no use prompting him. You must dictate to him.'

A Staff College invigilator c. 1906

Standards at Staff College before 1914 were low enough to justify the common soldier's view that, in the main, staff officers during the First World War were pampered dolts. During his final written examination the future Brigadier-General Sir James Edmonds tried to help one particularly obtuse candidate by hinting at some of the answers. The invigilator came up and Edmonds, expecting trouble, was amazed when the man made the above comment. The Invigilator then took over the task himself and dictated the answer to the slow officer.

Cromwell's 'Western design'

In 1654, at the end of the First Dutch War, Oliver Cromwell decided to put his redundant fleet to good use by seizing Spanish possessions in the Caribbean, the chief prize being the large island of Hispaniola. At the outset, England's Protector seemed to have an embarrassment of riches from which to choose his soldiers and sailors for the expedition. At this time Cromwell's regulars were some of the best soldiers in Europe and his naval commanders were the equal of the Dutch – the cream of the world's seamen. Yet the English expedition to Hispaniola is a strong contender for the title of the worst amphibious operation ever staged – and not only by England. Had the inmates of Bedlam assembled an expedition it could scarcely have exceeded in incompetence the one that left England under General Venables and Admiral Penn.

Cromwell's 'Western design' was conceived in a spirit of optimism. But from the outset it seemed to be dogged by every human folly and vice, leaving only the refuge of cynicism or despair for its survivors. Instead of top-quality English regular soldiers Cromwell found that he had to rely on the sweepings of the Portsmouth gutters. Regulars would not volunteer for fear that they were being hired out as mercenaries and lured away from their homes. Instead levies had to be raised by beat of drum in London and the south coast ports:

> . . . hectors, knights of the blade, with common cheats, thieves, cutpurses, and such like lewd persons, who had long time lived by sleight of hand, and dexterity of wit, and were now making a fair progress into Newgate, from whence they were to proceed towards Tyburne . . . but considering the dangerousness of that passage, very politicly directed their course another way, and became soldiers for the state.

As usual in the 17th century England was sending thieves, cut-throats and murderers to fight for her abroad. Perhaps this was not viewed as unusual; after all it was the officers who mattered most, and here Cromwell felt himself secure in his choice of Robert Venables and William Penn, men who had achieved renown in the English Civil War and afterwards in the wars in Ireland and against the Dutch. Admiral Penn (the father of the Quaker leader who founded Pennsylvania) had fought under Blake and Monck against such redoubtable foes as Tromp and de Ruyter. There was no doubting this man's abilities – yet was he the right man for a combined operation? His contempt for landsmen was well known and at times he could be tetchy and difficult to work with. A later diarist – no less a person than Pepys himself – called Penn 'as false a fellow as ever was born'. In mitigation it should be said that Penn at least arranged to get the expedition to the Caribbean and back; what happened in between was very much the fault of General Robert Venables. Venables' lamentable performance is one of the expedition's greatest mysteries. At 41, with a career of steady soldiering behind him, Venables seemed to be an ideal choice to lead the expedition. Yet Cromwell – usually a good judge of men – was wildly off the mark this time, though one cannot entirely dismiss the thought that the fault lay not with Penn or with Venables but with Cromwell himself. Not content with having a command divided in half, he compounded the fault by extending it to include three commissioners, in addition to the admiral and the general. Robbed of the centralized command that is often so vital in war, the 'Western design' was to be in the hands of a committee of five. One wonders how Cromwell himself would have felt at the battles of Worcester or Dunbar if his every decision had had to be confirmed by four other men. In two of the commissioners Cromwell was to be well served, but in the third – a drunken Irishman named Greg Butler – he had chosen a man little more than a buffoon when sober and a roaring devil when in his cups.

The first clash between Penn and Venables occurred before they had even left England. When Venables arrived at Portsmouth he found that the naval authorities there had ordered his troops straight onto the transports, with the result that he was unable to review them or even count them. As if that was not enough he found that

Oliver Cromwell, Lord Protector of England (1653–8). Cromwell's 'Western design' of 1654 was a case of too many cooks spoiling the broth. Confusion arising from the joint command of Admiral Penn and General Venables was compounded by the presence of three 'commissioners' sent to oversee their work.

there was to be a sharp division between the army and the navy in the use of the expedition's stores. The navy had done the loading, so Venables immediately accused Penn of taking the best for his own men. Venables may have been right for, once in the Caribbean, the soldiers never received many of the items that were supposed to have been loaded for them. But if Venables had grounds for complaint on that score his own personal behaviour was not above reproach. He had only recently married – for the second time and to a much younger woman – and insisted on taking his 'good lady' with him on the expedition. He explained that she might be of use as a nurse but while few of the officers begrudged him his conjugal bliss they did feel that Mrs Venables should not have involved herself so much in the general's affairs. As a result a joke circulated the fleet – 'petticoat over topsail'.

Once at sea and in his element, Admiral Penn was able to reflect on his good fortune, charged with an operation that could yield millions in booty and commanding fourteen ships of the line to accompany the transports and supply ships. It was the most powerful fleet ever sent to the Caribbean and should certainly overawe the Spanish defenders of Hispaniola. Cromwell had tried to keep the fleet's destination secret but as usual the enemy was quickly apprised of his plans. Nevertheless, the Spanish felt unable to respond to such a powerful expedition other than by sending out a new governor – the Count of Penalva – with a mere 200 musketeers.

Penn's seamanship was equal to all problems on the voyage and when the fleet arrived at their first port of call, the English colony of Barbados, it had suffered the loss of just one vessel, the *Great Charity*, which foundered in heavy weather. But now things began to go wrong. Barbados in the 17th century was not the island paradise of the 20th-century travel brochure. In those days it was regarded as the 'dunghill whereon England doth cast its rubbish'. The inhabitants included ex-pirates, and indentured servants and labourers transported as criminals from England. For some reason Venables was dissatisfied with the 2,500 soldiers he had brought with him, and the regiment of seaman raised by Vice-Admiral Goodson, and set about raising more troops among the colonists. This was a big mistake. The English commanders found themselves overwhelmed with volunteers to join their expedition. But they soon discovered that many were breaking their indenture and trying to escape from Barbados. By the time they left the island the English were equipped with 3,000 more recruits – to call them soldiers would be a travesty – the 'most prophane debauch'd persons that we ever saw, scorners of Religion, and indeed men kept so loose as not to be kept under discipline, and so cowardly as not to be made to fight'. Why Venables thought he needed these men is a mystery; the Spanish garrison on Hispaniola was very weak – numbering less than a thousand men – and would easily have succumbed to the regiments he had brought from England.

While on Barbados Venables discovered that the stores promised for his army were either substandard or had not been loaded in the first place. And there was so serious a shortage of weapons that a council of war considered abandoning the whole mission. In the end an appeal went out to the navy and to the planters of Barbados to make up the shortfall – an incredible 2,000 muskets, 600 pikes and large numbers of pistols, powder and ball. Either the authorities in England had sent the expedition off without weapons or else the navy had hidden them, intending to sell them at a later date. The situation was unprecedented: an army had been sent to war without any arms to its name. Venables had to shift for himself, calling up every blacksmith in Barbados to make the best weapons possible from the bits and pieces available. Although 2,500 pike heads were fashioned, there were no suitable staffs to be found, and 'cabbage stalks' had to be used, which were lightweight and easily snapped. Not without reason was the expedition described 'as the worst equipped ever to have left the shores of England'. But Venables hardly made things easier for himself by recruiting so many extra men on Barbados. There was scarcely enough food to feed his original 2,500, and once his numbers were doubled or even trebled, everyone faced starvation rations thereafter.

Meanwhile, Commissioner Butler had been sent to the island of St. Christopher. At that time the island was jointly owned by England and France: the centre belonged to England while both ends of the island belonged to France. Butler saw his mission as raising thousands more men from among the English settlers, but as England was temporarily at peace with France, Butler thought it prudent to make a courtesy visit to meet the French authorities. Having drunk most of Barbados dry, Butler now transferred his attentions to St. Christopher, arriving for his meeting with the French dignitaries in an alcoholic haze, falling from his horse and vomiting on the sandy ground. By the time Penn and Venables arrived at the island Butler had added 380 more undesirables to an English army that now numbered nearly 9,000.

While the English were busy recruiting an army numerous enough to invade Russia, the new Spanish governor, Penalva, was trying to make the best of his very limited resources on Hispaniola. On paper he had no chance at all against Penn's powerful ships-of-the-line. The town of Santo Domingo was close to the shore, well within range of the naval guns, and had a puny stone wall much supplemented by hedges. Penalva's only hope lay in the mistakes that the English might make.

Penalva cannot possibly have known how lucky he was in facing Penn and Venables. The innate irascibility of both men was exacerbated by the hot climate, with the result that minor disputes soon grew out of all proportion. How bad their relations had become can easily be seen by the fact that officers of both services felt it necessary to make a formal resolution pledging that neither would desert the other in its peril. As the fleet approached Hispaniola agreement was at last reached on a plan to capture Santo Domingo. It was perhaps more complicated than it need have been in view of the weakness of the Spanish garrison, but Admiral Penn refused the army's simple idea of an all-out assault in favour of a genuine amphibious operation involving two landings by soldiers and marines, combined with a covering bombardment by the navy. Venables seemed happy enough to go along with Penn's plan, but soon regretted his decision. While Penn stayed off Santo Domingo with his warships ready to begin the bombardment as soon as the soldiers were ashore, Venables was taken to the west of the town by Rear-Admiral Goodson to make a landing with the best of his troops near the mouth of a river, six miles from Santo Domingo. Meanwhile, to confuse the Spaniards, another smaller landing was to be made to the east of the town. But as Goodson and Venables sailed west they were unaware that they had left someone behind – the only pilot who knew where the landing was to take place. Goodson confidently sailed on, waiting to be told by the pilot where to anchor, but no message came. It was not until the ships had been travelling for some hours that the pilot's absence was discovered. By that time the fleet had sailed too far downwind, way beyond the intended landing beach, and a landing had to take place some 30 miles from Santo Domingo. Venables now had a two-day march to get to the town instead of a two-hour one, while the fleet swung at its anchor and the feint attack simply petered out in the torrid heat. As a demonstration of English lack of expertise in combined operations it was without parallel.

Venables and his men had been flung ashore on a hostile coast like a sack of mouldy vegetables. They had no maps of the area, no local guides, no tents, no tools – and useless weapons, which would snap on contact with the first real soldiers they met. They faced up to their 30-mile march in atrocious heat, not a few of them wishing that they had taken the shorter route to the scaffold back in England. At their first meal break it was discovered that their supplies consisted of mouldy bread, oversalted meat and virtually no water. Lacking water bottles, the men scooped up water from ditches with their hands, many succumbing almost instantly to 'the bloody flux'. Others, gorged on unripe fruit, had such disordered stomachs that they were forced to walk 'bare-arsed'.

The Spanish commander, meanwhile, had made the most of his limited resources and was offering the local inhabitants a bounty for the number of English heads they brought in. The local Black slaves and Creoles, known as 'cowkillers', had eagerly taken up the offer and lay in wait for Venables and his 'army' as they blundered through the countryside. The general, unaccustomed to warfare in the Caribbean, failed to use his cavalry to scout ahead and warn of ambushes. So, time after time the English were surprised by groups of 'cowkillers' wielding deadly lances, who took a heavy toll of the miserable English soldiery. When by chance Venables stumbled on a man who might have helped him – an Irishman who knew the area well and offered to guide the English column – Venables, who did not trust Irishmen, had him hanged.

An incident more typical of a schoolboy outing than a military operation tipped the scales even more against the English. Some of Venables' soldiers, foraging for food, found a statue of the Madonna, which they immediately declared was a 'Popish trumperie' and pelted with oranges, breaking it in pieces. This minor act of vandalism was observed by the locals, and the cry went up that the English had come as enemies of their religion. Venables was now facing a religious war to add to his troubles.

On the second day of the march panic broke out when a soldier was seen in the distance, apparently guarding the path down which the English column was advancing. Perhaps the Spaniards had at last come out to fight. Venables got his men into some kind of order and went forward in a gingerly fashion. But the soldier turned out to be English; he was part of the force which should have been landed to the east of the town. But this landing had also gone horribly wrong. The navy had been unable to find the right landing place and had instead put the men ashore to the west of Santo Domingo, exactly where Venables should have been landed two days before. Since he clearly could expect no help from the navy, Venables decided the best thing to do was to march to Santo Domingo and take it by storm. But he was reckoning without Commissioner Butler and his tricks. He had been in command of the force that should have landed to the east of the town but was instead now stranded to the west. Rather than going straight to Venables to explain that he was in the wrong place, he sent his men forward to attack Santo Domingo and walked straight into a Spanish ambush. Venables and his column had just caught up and the general was walking forward to inspect Butler's positions when all hell broke loose – with guns blazing and muskets firing – and the road was suddenly filled with English soldiers rushing back towards him. Venables lost his head, turned and ran, and took cover behind a large tree. Once the mob had passed Venables reappeared 'so possessed with terror that he could hardly speak'.

Although the English had lost just 20 men it was their self-respect that had taken the biggest drubbing. Venables found it impossible to face his men and went off in a sulk to the flagship. Claiming that he was going to consult Admiral Penn he ignored the admiral and spent seven days with his wife, while his army, demoralized and starving, clung to their ramshackle camp on the beach like hermit crabs. The food that had been brought from Barbados had, by an oversight on the part of the quartermaster, been taken from the store of condemned barrels, and as a result a number of men died from food poisoning. All the dogs and cats in the area were hunted like prize game birds, and the cavalrymen soon found themselves relegated to the ranks of footsloggers when their mounts found their way to the dinner table.

Sailing Master Whistler of the *Swiftsure* has left a vivid account of the English camp at Santo Domingo:

> General Venables, being aboard of our ship, and having a good ship under him and his wife to lie by his side, did not feel the hardship of the soldiers that did lie on the sand until the rain did wash it from under them, and having little or no victuals, and nothing to drink but water . . . and the abundance of fruit that they did eat, and lying in the rain did cause most of them to have the bloody-flux, and now their hearts were gone out of their doublets into their breeches, and was nothing but shitting, for they were in a very sad condition.

Admiral Penn did offer to send cooked meat and fresh water ashore from his ships to help the soldiers, but Venables would not hear of it. And when Penn offered to land sailors to storm the town for him Venables nearly climbed the mainmast in his rage.

Having regained his self-esteem in the company of his wife, Venables now ventured back to the beach with a new plan. He would take Santo Domingo by storm, and to do so he would advance down the main road to the town and fight it out in front of the walls. But this was playing into the Spaniards' hands, for Penalva had placed all his musketeers in trenches and behind hedges that lined the road. Had Venables sent anyone to the other side of the town he would have discovered that it was not protected by walls, but by hedges. The town was no more substantial than a Hollywood film set. But Venables was insistent that his men should give battle in front of the town walls as if they alone, and not just a flimsy line of privet, stood between him and victory.

At dawn on 25 April the assault force began to march towards Santo Domingo, led by Adjutant-General Jackson. Jackson was a baffling choice, since he was as incompetent a man as ever held rank in the British Army, and renowned for his keenness to save his own skin. He decided not to send out scouts to check for ambushes, and placed himself in the rear of the column, behind every other man in his command – even the cook.

It was a long walk down the dusty road – perhaps six miles – and the heat made men faint by the wayside. Suddenly the Spaniards burst from their cover with fierce yells and battlecries, while the English took to their heels, with Jackson – who showed a fair turn of speed for a man of his age – leading the flight. Two English officers drew their swords and threatened to kill anyone who played the coward and fled, but they were soon overrun. The best troops, whom Jackson had, oddly, placed at the rear, were swept away by those fleeing down the road towards them. Less than one man in fifty stood to face the Spaniards that day. Eventually the flight was stemmed when Venables – of all people – pushed his way through the stampeding English soldiers with a troop of sailors. By then the Spaniards had given up the pursuit, for Penalva had won a great victory without even trying. Nearly 400 English soldiers had been killed or crushed in the panic, including the only decent officers in the army. The 'cowkillers' were busy collecting heads and dreaming of spending their bonuses, while the English battle standards were paraded through the streets of Santo Domingo to derisive hoots and jeers.

Venables ordered a return to the beach. On the way Adjutant-General Jackson was seen seated under a tree having his back massaged by one of the women camp followers. Seeking a scapegoat for the disaster, Venables looked no further than Jackson. He was promptly court-martialled, had his sword broken over his head, and was reduced to swabbing the decks of one of the English ships.

The attack on Santo Domingo had failed ignominiously. There was nothing to be done but to sail away from Hispaniola in search of easier pickings. Relations between army and navy were now so bad that the soldiers refused to allow the troop of sailors to re-embark until all the soldiers were aboard, in case the fleet decided to sail away and abandon them on the beach.

But now the Spaniards blundered for the only time in the campaign. When they learned that the English were going to try their luck on Jamaica they failed to warn their compatriots, leaving the islanders vulnerable to a surprise attack by the English. But the island had its own way of dealing with the invaders. As the boats carrying Venables' men approached the shore near the mouth of the Rio Hayna, a rustling sound was heard coming from the dry reeds and undergrowth that fringed the beach. The men disembarked and moved cautiously towards the sound, only for the rustling to increase and become more widespread. Fearing that the reeds contained more of the dreaded 'cowkillers' who had so terrified them at Santo Domingo, the English soldiers turned and fled back across the sand to their boats and put out to sea, refusing to return in spite of the entreaties and orders of their officers. Later investigation revealed that the enemy that had put the English to flight was nothing more than hundreds of land crabs, which had been moving about among the reeds in alarm, thus producing the rustling effect that had so terrified the English.

Venables tried again but this time found himself confronted by a ragged militia of under 200 Spaniards, with just three cannon, commanded by what seemed at a distance to be an Egyptian mummy wearing a Spanish hat over its bandages. This was in fact the unfortunate Don Juan Ramirez de Orellana, who was afflicted from head to foot by a particularly unpleasant skin disease, and had to be carried everywhere in a litter. Such opponents were more to the liking of Venables' troops and soon the capital, Villa de la Varga, was in English hands.

On their return to England, after their 'triumph' in Jamaica, Penn and Venables were treated impartially by Oliver Cromwell – they were both clapped in irons and imprisoned in the Tower of London. The humiliation on Hispaniola had deeply angered England's Lord Protector. But his anger abated with time, and both men were released and allowed to retire into private life. Cromwell, after all, was as much to blame for the fiasco as anyone.

'I proceeded to become very annoyed and told the interpreter to go into the village again and to inform the King that if he did not come out forthwith and do homage to the British flag, I would set fire to the village and make war on him and his villagers.'

Lieutenant F.P. Crozier, West African Field Force, 1902

The Black Man's Burden – imperialism at its worst. This man was speaking for generations of British officers and gentlemen, who believed that it was their duty to introduce the poor, benighted savages of the world to the civilizing properties of the British flag.

'March or die'

The expedition to force Madagascar to accept a French protectorate in 1895 was the most disastrous colonial campaign of the Third Republic. The army won the contract to carry out the campaign by underbidding the Marines by nearly 30 million francs. As a result the planning was left to the War Ministry – which was accused of being totally ignorant of colonial affairs. Ignorance combined with cost-cutting by the army meant that the expedition was eventually deficient in most respects.

A committee was set up in Paris to plan the campaign and began by ignoring the experience gained from previous colonial campaigns. This was to be a new departure, bearing the imprint not of the past but of the latest thinking by the ministries of war, the navy, the colonies and the foreign office: it was to be a blueprint for the future. The committee estimated the strength of the enemy at 40,000 Hova tribesmen who, according to the latest scientific principles, could best be despatched from this world by a force of 12–15,000 French troops. The expedition's target would be to capture the city of Tananarive, situated high in the mountains of the interior, in the north of Madagascar. French forces would land at the port of Majunga on the west coast, which had a large harbour and was at the mouth of the Betsiboka River. This important waterway was navigable for 160 miles and would take the troops deep into Hova territory. Although estimates indicated that 20,000 porters and mule drivers would be needed to support the army on the rest of its journey, this number could be reduced to just 5,000 by the use of the *voiture Lefèbre* – a vast two-wheeled metal waggon weighing 500 pounds when assembled. Finally, the committee decided – controversially – that two-thirds of the troops should be white and drawn not from colonial garrisons but from France itself. Although these men might be unacclimatized and inexperienced in colonial warfare, their steadiness would more than make up for any deficiency.

Once the plan was made public experienced marine officers questioned whether it had been planned by the enemy or was a practical joke. The choice of Majunga as a port of entry was quite wrong, they protested. It was too far from Tananarive and was on the wrong side of the island anyway. The use of an east coast port, like Vatomandry, would reduce the marching distance to Tananarive by two-thirds. Secondly, why was it thought necessary to send as many as 15,000 men to fight the Hova? Five thousand lightly equipped colonial troops could have done the job in half the time.

But the planning committee was adamant. The Hova army was armed with European weapons and commanded by English officers. In addition, the terrain of the interior would enable the Hova to stage ambushes. In the eyes of the committee the journey to Tananarive might rival Napoleon's advance on Moscow. The hard-bitten marine officers shrugged and whistled down their pipes. They would wait their chance to say, 'I told you so'.

The French navy arrived off Majunga on 15 January 1895 and proceeded to flatten the small port, which was largely made up of wooden houses shaded by mango trees. It had been an idyllic and peaceful spot. The first Frenchman ashore, the previous night, had been Captain Emile Reibell, and he had soon discovered two enormous drawbacks to landing at Majunga. Trying to snatch some sleep in a deckchair on the beach he had found himself exposed to the secret weapon of the Hova – mosquitoes, and disease-bearing mosquitoes at that. His second discovery was even worse. Majunga was the worst possible place on God's earth to land an army. It may have looked the perfect spot for a disembarkation when viewed on a map in Paris over wine and foie gras – but out there on the fever-ridden coast of Madagascar it looked perfectly awful. Far from offering a clear run in to sandy beaches, the harbour was in fact surrounded by a coral reef which prevented any large ships from even approaching the shore. As if this was not bad enough the huge bay was subject to a dangerous swell that would capsize all the rivercraft that the French had brought with them. There would now be no comfortable river journey into the interior. The troops would have to walk – three times further than if they had landed on the east coast.

The French commander, General Duchesne, next addressed the problem of

French troops burn the bodies of Hova tribesmen killed in action in Madagascar, 1895. Deficient in almost every respect – and ruinously expensive – the French expedition to Madagascar took a heavy toll both of French troops and of the indigenous Hova people.

porters. Five thousand had been the estimate in Paris, but Duchesne soon found that porters were scarcer than brain cells in the French War Ministry. Efforts were made to bring in men from as far away as Indochina, and some were brought down the coast from Somalia. Algeria supplied 3,500, but even so the French still had less than one porter for every three soldiers. Only the *voitures Lefèbre* could save the day. But the vehicles had been planned for use in the Madagascan highlands, not in the tropical swamps of the coast, where they simply sank into the soft ground. Duchesne was not deterred. 'We have no road,' he said. '*Eh bien* – we build one.'

While the French dug themselves graves on the coast, the Hova army assembled outside Tananarive. 'They were a ragged lot, and [of] discipline there appeared to be none,' wrote an English observer. The Hova were excellent at simulating warfare, performing fierce dances and miming and gesturing with gusto. Whether or not they would stand up to an enemy was entirely another matter. A few of the royal guards were equipped with Remington rifles, some had antiquated Snyders, with few bullets, while most of the Hova were armed with muskets or bows and arrows.

Morale in the Hova camp was high after they heard that General Duchesne was building a road from the coast. They knew that if the French stayed there long they would all die of fever. As one of the Hova said, 'We were at a loss to understand how any sane general could keep his force so long in the deadly lowlands instead of pushing on, without delay, at any cost, to the healthy highlands, which were but a few days march distant.' Instead the road crept inland like a blind worm, a few feet every day, along the bank of the Betsiboka River. The *voiture Lefèbre* was imposing the unbearable burden of modern technology on an environment unchanged since the dawn of the world. Hundreds of soldiers died, first of disease and later, as the supply system broke down, of malnutrition and starvation. A French officer wrote, 'All day our men make crosses, dig graves, and every evening it is always the same funereal and lugubrious alignment of white cadavers.' Malaria and dysentery ravaged the French camp. Another officer wrote, 'We can no longer count the victims. And for what? To drag behind us the Lefèbre waggons. Whoever decided to send them to Madagascar is a real murderer. The cemeteries are filling up. When are we going to march forward?'

But once started, the obstinate Duchesne would not stop until the road was built. It never occurred to him to move the rest of his men back to the coast while the work went on or to send on a lightly equipped column against the Hova. He was a veteran of France's terrible wars in Tonkin and Formosa and believed men were born to suffer. As he said, 'When you go on campaign, you don't go to a banquet.' It was a banquet for the mosquitoes, however, as entire units of French troops were wiped out or lost 75 per cent of their number to malaria. Duchesne had already suffered a major defeat without even seeing an enemy soldier. Once when he saw ten soldiers led by three officers returning from a day's roadbuilding he remarked, 'That's a lot of officers for so few men'. He was told that the thirteen were all that remained of an entire company of 250 engineers.

After seven months of backbreaking toil even Duchesne realized he was never going to complete the road. All that was left was to pick the best men he had left and to send a light column inland – a course of action that had been suggested by marine officers back in Paris before the expedition began. Eventually the best 1,500 men, including a detachment from the Foreign Legion, were assembled to make the march inland. They were not an impressive sight, looking . . .

> . . . so dejected, so depressed, so pale, that one would have believed them more dead than alive. Their clothes were in rags, their boots in pieces, their helmets, too large for their emaciated faces, fell to their shoulders, covering almost entirely their yellow faces where only eyes the colour of fever seemed to exist. And they seemed so pathetic, so poor, so miserable, that unconsciously tears came to the eyes.

On 14 September the march inland began. Already cynics were calling it 'the suicide of General Duchesne'. Soon the French soldiers were climbing through wooded and mountainous terrain. The air was fresh and clean but the Hova used the rocks to stage a dozen ambushes. Duchesne was cautious, being uncertain of the quality of the Hova as fighters. In Paris the planners had painted a picture of a semi-European style army, led by English officers. But as usual the planners were wrong. The quality of the Hova army had been wildly overestimated. When Duchesne's men approached the first Hova entrenchments the tribesmen simply threw down their weapons and ran away without a second thought. The relieved Frenchmen 'saluted this grotesque flight with the most energetic catcalls'. The martyrdom of the coast had been unnecessary – the campaign was a walkover.

At the first sight of French troops the Malagasy commander had told the Hovan prime minister, 'I can do nothing. My men will not stand. They run away as soon as they perceive that two or three of their friends have been killed. Nothing will stop them.' In fact the Hova ran away with such gusto that on one occasion 300 of them mistakenly ran off the edge of a precipice and were killed. In just ten days the Hova army had been reduced by desertion from 7,000 to just 1,313. Most of them just fired off their ammunition into the air and then fled saying they had no bullets left. The last defence of the capital Tananarive was to be made by local peasants who had been rounded up, shackled together and driven like slaves into the city.

The French were soon aware that the Hova soldier was a contradiction in terms. Yet the French column was so exhausted by the climb after its enervating stay in the swamps that it hardly had the strength to continue the march. Ten per cent of the French troops had been lost since the march began, many through exhaustion but others through suicide. On one day six legionnaires killed themselves rather than continue marching. The psychologically fragile legionary soldiers were better when working or fighting – marching gave them too much time to think. One after the other, French soldiers killed themselves, by hanging or shooting. One shot himself in the leg in the hope that he would have to be carried by others, but his ploy failed when he died of blood poisoning. The French force was thoroughly demoralized. Those who fell by the wayside were murdered by the local Hova peasants; the women took a peculiar delight in torturing their victims and parading their private parts through their villages. By this time, however, the Hova army was the least of Duchesne's problems. His march had become simply an exercise in survival, with many of the French barefoot and dressed in rags. Their deplorable state should have inspired the remaining Hova to defend their capital. An English officer recorded how they did it.

> Bodies of [Hova warriors] were marched up hills and then marched down again. I saw 1,000 Betsileo spearmen rush up a height at the back of the hospital; having reached the summit they waved their spears and raised a great shout; and then they quietly came down again, soon to recommence the same performance on some other height . . . As soon as reality approached, as soon as the defenders found themselves within range of the French shells, and even before that, they bolted to some other position, where they could make another demonstration of battle without incurring any personal risk. It was not war, and it was not magnificent.

Duchesne ordered his artillery to open fire on Tananarive. One of the first shots knocked off the corner of the Queen's palace. This seemed to knock the stuffing out of the Hova resistance, and a white flag was promptly waved from a flagpole. The French had won – but there was to be no triumphal procession into the capital. An English officer reported that the French resembled the walking dead, one being covered from head to foot with a mass of flies, as if he was already decomposing within his uniform. Even at the moment of victory Madagascar had a final trick to play on Duchesne's long-suffering warriors. As the French formed up in column to move forward a herd of pigs suddenly charged out of Tananarive, skittling men left, right and centre and inflicting more casualties than the whole of the Hova army.

'Machine guns have not, in my opinion, much future in a campaign against a modern army.'

General John Adye, writing of machine guns in 1894

Britain's experience of colonial wars in Africa and Asia often blinded her generals to military developments elsewhere, notably in Europe and the United States.

The Madagascan campaign had been doomed from the moment planning first began in Paris. The committee got everything wrong, from the choice of port to the mode of transport. Cooperation between the various departments was poor; for example, many French soldiers were denied quinine in Madagascar simply because the medical supplies had been loaded in the naval transports first and therefore could not be unloaded until everything else had been removed. Duchesne – a tough commander – was perhaps too tough for his own good. A more compassionate general would have looked for ways to save the lives of his men even if this meant seeking more subtle ways of overcoming a Hova enemy whose quality had been exaggerated. Deaths from disease were well over a third of all combatant deaths, mostly from malaria but also through typhoid and dysentery, resulting from poor hygiene levels in French camps. A thousand men died on the return trip to France alone because of execrable conditions in the troop transports. Quite apart from the heavy toll exacted by disease, the number of suicides exceeded the battle casualties. One battalion from the Foreign Legion lost 104 men to suicide and just 34 in battle. The savage instruction to weary Foreign Legionnaires – 'March or die' – was never more clearly illustrated than in this truly hellish campaign.

The road to Ballynamuk

The most uncombined of all combined operations must have been the French attempt to invade Ireland in support of the patriot Irishman Wolfe Tone in 1798. Napoleon himself had considered leading the expedition, but had finally decided to go to Egypt instead, preferring to play the part of Alexander the Great than tramp through peat bogs.

The French invasion force was divided into two parts: one, led by General Amable Humbert, with 1,000 French soldiers, was to sail from the port of Rochefort, while the second and larger section, some 3,000 men under General Jean Hardy and accompanied by Wolfe Tone himself, would leave from Brest. The departure of the two flotillas was supposed to be synchronized, with a rendezvous off Donegal, in the north of Ireland, planned for later. It was a simple plan, but it made little allowance for the weather. By August 1798 Humbert's force was ready – Hardy's was not. So Humbert sailed anyway and instead of going to Donegal went instead to Killala, in County Mayo.

The first major French invasion since 1066 began in a curiously subdued way. Instead of French troops storming ashore like latter-day Vikings, raping, pillaging and burning, General Humbert opened the front gate of a neat Georgian house, walked up the pebbly drive and knocked on the front door. It was opened by a servant and at once the master was called. Humbert introduced himself and told the old gentleman – who turned out to be the Protestant bishop, Dr John Stock – that he was the commander of a French army that had just invaded Ireland, with the intention of helping the people win their independence from Britain. Dr Stock thanked Humbert and asked to be excused as he had a sermon to complete.

Humbert now marched his men into Killala and hoisted a green flag decorated with a harp and bearing the motto 'Ireland Forever'. A crowd began to gather. The public houses turned out and soon Humbert had recruited a thousand volunteers, enthusiastic to fight the English and take the French gold. But the Irish had no idea what they were letting themselves in for. Humbert marched his army around the country lanes near Killala, attracting more recruits, and wondering when General Hardy was going to show up with the main force. But Hardy's fleet was kept in harbour by contrary winds – to all intents and purposes Humbert was on his own. Humbert himself soon reached the same conclusion and marched out of Killala towards Castlebar, the county town of Mayo.

News of the French invasion had quickly reached Dublin and the viceroy there – General Cornwallis (who had fought in the American War of Independence) –

The Irish patriot, Theobald Wolfe Tone, was to have been installed in power by the French invasion of Ireland of 1798. In the event the invasion was an ignominious – even comical – failure, and Tone died by his own hand in prison after being captured by the British at sea.

began to assemble his forces. In Mayo, General Hely-Hutchinson had occupied Castlebar with a force of 4,000 regulars, equipped with artillery. On paper this army looked quite capable of taking care of Humbert's rabble. But on 27 August – exhausted by their long march and outnumbered three to one – Humbert's troops, helped by his Irish volunteers, put the British to flight, inflicting over 400 casualties and capturing nine British guns. The performance of the British regulars was quite remarkable. They ran so fast that the battle was ever afterwards known as the 'Castlebar Races'. Some infantry retreated 30 miles to Tuam in County Galway, while the cavalry fled as far as Athlone – 63 miles in 24 hours.

Having taken Castlebar Humbert was at a loss what to do next. Where were Hardy and Wolfe Tone? Was he expected to conquer Ireland all on his own, with so few men? In the absence of any other orders, he would give it a try. With just 800 men and 1,400 Irish volunteers, Humbert set out to march to Dublin, hoping to raise the countryside as he passed. But unknown to Humbert, Cornwallis had left the capital and was closing in on him with 40,000 regular troops. When the French reached Ballynamuck in County Longford they found themselves facing an army drawn up for battle. Although outnumbered 20 to one, Humbert was not dismayed (even if he was, it would not have done to show it) and ordered his men into the attack. For 30 minutes Humbert's men fought a *baroud d'honneur* before surrendering to the English general. It had all been a 'jolly good show' and Humbert was generously entertained by the English officers with tea and cakes, while his men were patted on the back and taken away to be paroled. Humbert's Irish volunteers enjoyed a different fate: they were massacred by General Lake's dragoons.

While Humbert's invasion ended in tragic farce in Ballynamuk, the second part of the combined operation finally got started. On 16 September Hardy and Wolfe Tone left Brest only to run into a powerful British naval squadron off the coast of Ireland. After a fierce ten-hour battle the French ships were scattered and Wolfe Tone captured. He was afterwards sent to Dublin in chains. Before he could be publicly executed Tone cut his throat in prison. Thus ended the second part of the French invasion of Ireland.

But even this was not the absolute end. A third part of the combined operation – one that nobody had even told Humbert to expect – had set sail from Dunkirk. Another Irish patriot, Napper Tandy, sailed into Donegal harbour with 270 French soldiers on the same day that Hardy was setting out from Brest. Tandy went cautiously ashore, met an old drinking companion of his, learned that Humbert had been defeated, spent the night drinking Donegal dry, was carried back on board a French ship, and escaped unnoticed back to France. It had been a 'hit and run' operation and Tandy had made it back into exile.

The fourth part of the combined operation – a part that neither Hardy nor Humbert had known about – set sail from Rochefort on 12 October, commanded by a Captain Savary. With the same ships that had transported Humbert in the first place, Savary took with him the entire band of the 70th Regiment. He sailed into Killala harbour, came as close to the shore as he dared, ordered the band to play, and then directed the fleet to return to France. No British response was recorded but the locals apparently liked the music. Dr John Stock was momentarily distracted from his sermon. Thus ended the silliest combined operation known to history.

'Thank God! Then I'll be at them with the bayonet.'

General Sir Hugh Gough at the battle of Sobraon, February 1846, on being told that the British were running short of ammunition

Gough was typical of many early 19th-century British commanders, brave but stupid. He was a butcher who squandered British lives by his absurd belief in frontal assaults and the use of the bayonet.

A Norwegian fiasco

Seniority in the armed forces can be a touchy subject, particularly where an officer senior in rank is asked to waive his seniority and defer to someone junior in rank. This can occur in combined operations – one has only to remember Lieutenant-General Bryan Mahon during the landings at Suvla Bay in 1915 who, stripped of nine of his twelve battalions, refused to operate as a mere brigadier and tried to resign in the middle of the battle.

The German invasion of Norway on 9 April 1940 caught the British government if not by surprise at least in a state of chaos. Britain had already assembled forces to intervene in Norway and now found that her move had been pre-empted by the Germans. The campaign that followed – aimed at freeing Norway from the Germans – was one of the most inept combined operations ever carried out by Britain. It might have helped if the Admiralty and the War Ministry had at least been on speaking terms. How a combined operation was to be effective when the naval and military commanders did not know each other's orders and were pursuing virtually contradictory ends is hard to imagine. To appoint a naval commander as senior to the military one by as many ranks as separate a private from a commissioned officer was asking for trouble. Why an admiral of the fleet – senior to every naval officer on active service, as well as to the commander-in-chief of the Home Fleet – was chosen to command such an operation is difficult to fathom, though he was of course a particular favourite of the First Lord of the Admiralty. In any case, it cannot fail to have made the military commander, Major-General Mackesy, extremely uncomfortable.

The decision to transfer the troops earmarked for Stavanger, Bergen and Trondheim to Narvik may have seemed a good idea at the time but was in fact futile. These men had been loaded with a view to landing in a friendly country and their stores had been shipped with no consideration of tactical usefulness. How they could suddenly transform themselves into a force fit to make an opposed landing, when none of them knew where to find vital weaponry and supplies, was never explained at the time. Reacting to the rumour that handfuls of Norwegian troops were still holding out around Narvik, Britain decided to rush to their aid. On 11 April Mackesy sailed from the Clyde with some of his troops and with the naval commander's chief of staff, but not the admiral himself, who was still in London receiving a *viva voce* from Churchill. At this stage neither Mackesy nor Admiral of the Fleet the Earl of Cork and Orrery had met to discuss plans. Neither had the Admiralty and the War Office concerted the orders given to their commanders. It was to be an uncombined operation in the best tradition of the Dardanelles and earlier fiascoes.

The 2,000 German troops at Narvik were in a desperate position, short of weapons and ammunition – their supply ship having been sunk by British destroyers – and dug into the snowy heights behind the town. But Mackesy saw things differently. He decided that a major operation, involving Norwegian troops and British and French reinforcements, would be needed to prise the Germans out of Narvik. This attitude came as a shock to the Earl of Cork who, inspired by Churchill's

Admiral of the Fleet Lord Cork and Orrery (centre) flanked by Vice-Admiral John Cunningham (left) and First Sea Lord Sir Dudley Pound (right). Cork was designated naval commander of the expedition to Narvik in 1940, but he failed to coordinate his activities with the expedition's military commander, Major-General Mackesy.

'gung-ho' advice, favoured an all-out attack on the town. He knew nothing of Mackesy's orders, only that he was already on his way to Narvik with troops. While chasing to catch up with Mackesy, Cork received an important but misleading signal from Vice-Admiral Whitworth in the battleship *Warspite*, reporting the defeat of the German destroyers and his opinion that Narvik could be taken with little resistance. According to Whitworth's rather optimistic report, the Germans were streaming away into the woods, apparently thoroughly demoralized. This was just what Cork wanted to hear, and he tried to contact Mackesy to arrange a meeting at Narvik. Instead he learned that the general was waiting for him at Hardstad, 30 miles away from Narvik, and received a further signal from the Admiralty telling him that he must act only in conjunction with Mackesy and not on his own initiative.

On 15 April Mackesy and Cork met for the first time at Harstad and soon realized that their orders and their own interpretations of them were so different that combined action could almost be ruled out straightaway. Cork was convinced that an immediate landing, preceded by a massive naval bombardment from the *Warspite*, was possible. Mackesy did not agree. He thought that without proper landing craft it would only be possible to land 400 men at a time and that this would lead to heavy casualties. In addition the naval bombardment would inflict unacceptable civilian casualties and his brief was to avoid those if at all possible. Mackesy then produced a bombshell. He told Cork that his orders from the War Office specifically forbade him to land in the face of any opposition. Thus, if we take Mackesy at his word, there seems to have been little reason for him to have gone to Norway in the first place. However, Mackesy was being less than straightforward with Cork, as a letter to Mackesy from the Chief of the Imperial General Staff makes clear. It told him to make use of naval action if possible, and added that 'Boldness is required'. So Mackesy was blaming his orders for what was in reality a personal lack of boldness. Yet it is wrong to apportion too much of the blame to Mackesy. His mission had originally been to make a peaceful landing, among welcoming crowds of Norwegians, not to fight his way ashore against crack German mountain troops. His men were unprepared for such a campaign, lacked proper winter equipment and training and had no proper landing craft. Boldness was one thing, irresponsibility entirely another.

The Admiralty naturally blamed the War Office for not keeping their man informed of Mackesy's orders. But when the War Office responded by asking to see Cork's it soon became clear that the Earl had no written orders at all. Everything had been done by word of mouth between Churchill and his commander. It was an odd way to run a war.

On 20 April it was decided to put Cork in charge of the whole operation, with Mackesy subordinate to him. But the damage had already been done and the British had missed their chance five days earlier. On 24 April a naval bombardment took place but with disappointing results. The Germans had recovered from their fright and had dug well in. Cork, meanwhile, had earmarked 8 May for a full-scale seaborne landing and had received landing craft from Britain. But again Mackesy was unhappy and made his views clear. Cork naturally overruled him but it soon became apparent that all the military men present agreed with the general. Cork felt he had no choice now but to drop the idea. A new plan to land troops at the head of a fjord ten miles north of Narvik was put into action by recently arrived French and Polish troops, and a new ground commander, Lieutenant-General Claude Auchinleck, arrived to replace the discredited Mackesy. But already events in France were overtaking the Norwegian campaign. Although on 28 May Narvik was at last taken, this could hardly compare in its effects with the shattering defeat of the Anglo-French armies in France and the evacuation from Dunkirk.

Breaking French windows with guineas

During William III's struggle against the France of Louis XIV, in particular during the Nine Years War (1688–97), the English could rarely steel themselves to endure the expense and casualties of a sustained land war on the continent. Geography – as always – gave them a unique advantage which they proceeded to squander time and time again. Unwilling to commit themselves entirely to either a continental or a Mediterranean 'blue water' strategy, the English preferred to invest their money in 'descents' on the French coast, or on French possessions in the Mediterranean or the West Indies. Such operations made use of England's traditional strengths, naval rather than military, and could also be controlled by the English themselves, rather than by Dutch or German generals in Flanders. William III, with his Dutch upbringing, never failed to regard the English as amateurs and pretenders to the military art. As he liked to point out to his English advisers, France suffered as little damage from these English raids as Louis XIV did from stubbing his toe. It was not until England was prepared – in the War of the Spanish Succession (1701–13) – to send troops and commanders of note to the continent that French hegemony in Europe would finally be broken. In the meantime the English were content to paddle in the shallows rather than risk themselves in the deep water of European warfare.

The naval victory of La Hogue in 1692 gave England supremacy in the Channel, and opened the way for a combined operation against the French port of Brest. The operation's prime aim was to force the French to withdraw troops from the main battlefields in Flanders, Germany and Italy to deal with the English threat to their west coast. England saw it as a way of helping her allies without risking too much equipment and manpower. It all seemed so easy that everyone took it for granted that an expedition against Brest needed little planning. All the navy had to do was carry the troops there and then row them to the beach, after which they would be on their own.

Preparations for the Brest operation began in 1693. Troops were recruited and billeted in Hampshire and Sussex while transport vessels were made ready. But it was not until the spring of 1694 that the scheme got the 'go-ahead' from the government, led by Whigs including Lord Cutts, the Earl of Macclesfield and the eventual commander of the expedition, Thomas Talmash. Whigs such as Talmash, and his colonels Thomas Erle, Richard Coote, Samuel Venner and Henry Rowe, supported an assault on Brest for political and economic reasons, hoping that it would reduce attacks on British ships in the Bay of Biscay. William III – dependent on the Whigs in Parliament for funding his war with France – was forced to support the Brest expedition as a fact of political life. But there was another, parochial reason behind the Brest expedition. The English army was tired of playing second fiddle to William's Dutch and German generals. It was time to launch an English operation, planned, equipped, officered and commanded by Englishmen. William could not easily resist the patriotic pressures to place Lieutenant-General Thomas Talmash at the head of a purely English operation. Ironically the choice of Talmash irritated not the Dutch and Germans so much as Talmash's great rival, John Churchill, first Earl – and later first Duke – of Marlborough.

The orders that Talmash received before sailing for Brest were very vague. From the start he never knew whether his task was to destroy the shipyards there, burn the ships and generally wreck the port, or whether he was to create diversions up and down the coast in order to force the French to withdraw troops from other theatres of the fighting. Eventually he would have to choose or have the choice made for him by the enemy. For the French were quite aware that he was coming. And the way in which they found out reflects no credit on England's greatest soldier – John

Churchill. Quite simply Churchill, for personal and political reasons, had written the following letter to the King of France:

> It is only today that I have learned the news I now write to you; which is, that the bomb-ketches and the twelve regiments encamped at Portsmouth, with the two regiments of marines, all commanded by Talmash, are destined for burning the harbour of Brest, and destroying all the men-of-war which are there. This will be of great advantage to England.

Churchill was putting personal interests before those of the state, trying to secure Talmash's failure and disgrace so that he could replace him as commander-in-chief. Oddly, he was not the only traitor in the English camp. Lord Godolphin, a Tory politician, had also written to tell the French, as had the Earl of Middleton. Louis had responded by ordering his great engineer Vauban to fortify the defences around Brest. Within days Vauban had mounted 90 mortars and 300 additional cannon, as well as rushing in a further 4,000 troops. Brest was now bristling with defences and Talmash was walking into a trap.

While his colleagues tried to make certain that Talmash failed, the navy betrayed such rank stupidity that it is impossible to believe that they were not out to ruin the expedition as well. In case the French were in any doubt as to Talmash's destination, and the point at which he intended to land, their Lordships sent three frigates into Camaret Bay to attack a French convoy there. Vauban immediately assumed that they had come to the bay to spy out landing beaches, and intensified his work on the defences there. So fantastic were the developments at Brest that London newspapers carried a daily 'Letter from Brest' giving readers the latest information on the growth of French armaments there. On 4 June the letter reported that there were now 400 cannon surrounding the port, as well as 9,000 French regulars. If it had been the government's intention to draw French troops away from other areas, they might have concluded that they had done well enough already and not bothered to send the expedition, which was compromised already and doomed to fail. Yet no attempt was made to call off the mission or alter the plans. With his political cronies Cutts and Macclesfield as his major-generals, Talmash left England at the end of May with 6,000 men. He was not feeling well and wrote to his brother that he 'was engaged more than ever, I must say, much against my will'. As if hypnotized, the English were following their leader into a trap of their own making.

On 7 June the English fleet sailed into Camaret Bay and anchored. The French had not actually arranged a welcoming committee, with bunting and bands playing, but the reception was not far off it. No sooner had the English ships arrived than two huge mortars opened fire, hurling shot two and a half miles out to sea. It was now obvious to Talmash that he could no longer rely on the warships closing in to support his landings. A single lucky shot from one of the mortars would have gone straight through the decks and the keel of a man-of-war.

That afternoon Talmash, having decided to take a look at the landing beaches himself, was rowed towards the shore with two senior officers. He found no signs of French soldiers or any entrenchments. Perhaps the newspapers had been mistaken – or perhaps it had all been French propaganda. In any case, Talmash felt confident

The military engineer and marshal, Sebastien Le Prestre de Vauban, was one of the foremost French commanders of the 17th century. His strengthening of the defences around the port of Brest – at the behest of Louis XIV – made the town virtually impregnable and condemned the English raid of 1694 to certain failure.

enough to tell a council of war that there was nothing to prevent a landing the next day.

The morning of 8 June was foggy and at 7 o'clock the signal was given for the first wave of troops under Cutts to make for the shore. It had been arranged that the landing craft should travel in an echelon, but soon all was confusion as some rowed faster than others. Half-way across the bay an effort was made to get the boats back into some sort of order but all this did was to give the French defenders more time to get ready. Soon the fog lifted and the bright sunlight revealed an amazing sight. The French gunners could hardly believe their eyes; the bay was covered with small unprotected boats, filled with English soldiers. As the landing craft drew near to the shore they were hit by a hail of musket fire and chain-shot, cannister and cannon balls which caused many of the boats to tip over. The beach Talmash had chosen – and had thought undefended – was just 300 yards long and was clearly protected by three lines of trenches, each filled with 150 musketeers. In addition, batteries of cannon now poked out their black stubby barrels. Cutts, leading the first wave, seemed uncertain whether to go on or fall back. In the meantime his men were being cut down all around him. Talmash, in a fury, had himself rowed over to Cutts, calling 'My Lord, is this following of orders? Do you see how the boats are in disorder? Pray, my Lord, let us land in as good order as we can.' Cutts responded by sending a few men onto the beach, but did not venture there himself. Talmash however was made of sterner stuff, and strode through the waves and onto the sand. Backed by three officers and just nine grenadiers, he rushed forward to the cover of some rocks. When he called for more support he was joined by a further 200 grenadiers. But they had walked into a death trap. French swordsmen rushed forward and drove the grenadiers back to their boats, leaving Talmash stranded virtually alone. When French cavalry charged down the beach, the remaining English grenadiers threw down their weapons and surrendered. Talmash was shot through the thigh and had to be half-carried, half-dragged back into the sea, but his boat had been abandoned by its crew. Only when Lord Berkeley offered some sailors a reward of £5 was a boat sent in to rescue the commander. With the assault in tatters, Macclesfield took the only decision possible and ordered the landing craft to make speed back to the English fleet.

Talmash was treated aboard the *Dreadnought*. Although his wound should not have proved fatal the surgeons 'did for him' and he died when gangrene set in. The expedition, which had lost 300 men, returned to Plymouth to find the English press incensed and looking for scapegoats. The search proved to be an easy one. The unfortunate Talmash, unable to defend himself from the silence of his coffin, provided the perfect scapegoat It was soon accepted that Talmash had made every mistake possible. Whether Marlborough smiled when he heard of the posthumous disgrace of his brave rival is not recorded. He was going to live long enough to become England's – and Europe's – greatest general while Talmash was food for worms.

England's attempt to 'break French windows with guineas' had ended in farce. Louis XIV – tongue in cheek – had a medal struck to commemorate 'the massacre of the English and Dutch on the shores of Brittany'. So amateurish had the English attempt been that it proved a death sentence for the strategy of 'descents'. From now on the English would have to face the French in the context of the kind of warfare they hated – the bloody land battles and sieges of Flanders. Talmash had shown more courage than sense. Under a different – albeit less immediately attractive – leader the English would recover their pride and break the French domination of Europe at Blenheim, Ramillies, Oudenarde and Malplaquet. Ironically on these famous fields they would be led by the man whose treachery had undone Talmash – John Churchill, first Duke of Marlborough.

'Staff College officers! I know these Staff College officers! They are very ugly officers and very dirty officers!'

The Duke of Cambridge, 1886

It must have been difficult to keep a straight face when the Duke of Cambridge was sounding off on his favourite subject, the fossilization of the British Army – a process to which he had contributed in no small measure. The Duke regarded the Staff College as a threat to the 'sublime ignorance' which had for so long carried Britain through her wars.

CHAPTER 3: WAR IN THE AIR

The bomber will always get through

Some have argued that the strategic bombing campaign by RAF Bomber Command during the Second World War turned British servicemen from warriors – fighting for their country against the armed forces of an enemy – into terrorists, targeting civilians and deliberately spreading terror as a means of breaking enemy morale. It may be claimed that the naval blockade of Germany between 1914 and 1918 started the process, and that both the blockade and the civilian bombing were merely facets of 'total war'. It is a moot point. Yet the saturation bombing of Germany's cities – with the Hamburg firestorm and the destruction of Dresden in 1945 standing out as particular horrors – may have been not only unethical but also completely pointless. As the saturation bombing failed in its purpose – neither breaking enemy morale nor destroying Germany's war industry – then it must be classed as one of the biggest strategic blunders of the entire war. Thousands of lives were lost over Germany, apparently to no good purpose. Huge bombers, having transported tons of explosive hundreds of miles across Europe, often ended up by dropping their bombs harmlessly in fields. Having done so, many of them never returned, falling from the sky like debris from a junkyard. The labour of thousands of workers in factories all over Britain was scattered across northern Europe and mostly to no good purpose. Raw materials that could have been better used to build tanks, or ground-support aircraft, or night fighters, or dive-bombers, or machine-guns, or rifles, or iron bedsteads was wasted in fulfilment of an obsession – the obsession with the bomber as a war-winning weapon. Far from winning the war for Britain, the bomber almost lost it by absorbing the resources that should have been used in the 1930s to equip Britain with enough quality fighters to defend her shores against the German threat.

The development of heavy bombers at the end of the First World War, combined with the civilian casualties suffered in the Gotha raids on London in 1917 and 1918, contributed to a dread of aerial bombing out of all proportion to its actual threat. After 1920 films were made and books were written emphasizing the horrors of this new and terrible form of warfare against which there was no defence. In 1921 Hugh Trenchard declared that 'the next war could be won by bombing alone, by destroying the enemy's will to resist'. These were potent words, and from them developed a whole strategic philosophy that was to dominate British – and later American – air chiefs and lead them up a blind alley. Winston Churchill entered this alley quite willingly in 1940, for lack of any other way to go, but he was wise enough to see that it led nowhere and he turned back in time. But for Arthur 'Bomber' Harris there was no turning back, and hundreds of thousands of Allied airmen and German civilians paid the price for his obsession.

In 1921 Trenchard was only echoing the words of other prominent airmen of the time, including Italian General Giulio Douhet, who claimed that the next war would last only a matter of days, during which time bombers would inflict millions of civilian casualties. No power on earth could save the civilian populations: the bomber would always get through. In the House of Commons Stanley Baldwin said, 'The only defence is offence, which means that you have to kill more women and children more quickly than the enemy if you want to save yourselves.' Baldwin's words were wrong, even in 1921. It is strange how able men failed to see that technical developments in one area of aerial warfare – bombing – would be matched by developments in others such as anti-aircraft defence or fighters. It was as if their capacity to think rationally became atrophied by fear of what they perceived that bombers could do. A Luftwaffe general commented in 1944 that Douhet's belief that bombing would shatter civilian morale was a reflection of the Italian national character, in that he knew how Italians would react to massed bombing and assumed everyone else would do the same. So dangerous was this petrifaction of intellect in Britain that the need for fighter planes was overlooked during the 1930s until it was almost too late. Trenchard was even on record as saying, 'Fighter defence must be kept to the smallest possible number.' And when war did break out in 1939, as much

Air Marshal Sir Arthur Harris perusing photographs of bomb damage caused to German cities by one of Britain's strategic bombing raids. 'Bomber' Harris, like many of his contemporaries, believed that German civilian morale could be destroyed by a campaign of strategic bombing, but the evidence has been far from conclusive.

labour and capital went into the production of heavy bombers as was allotted to the production of equipment for the whole of the British Army. This was a philosophy of faint hearts who feared to face the German soldier on the battlefield, but were prepared to drop bombs indiscriminately on his family from a safe height of ten thousand feet. It was an attempt by British leaders, shocked by the slaughter of a 'whole generation' on the Western Front, to find an indirect way of winning a war.

The false notion that the bomber would always get through was a product of First World War thinking. Then anti-aircraft and fighter defence had been in their infancy and the effect that the few bomber raids had on public opinion was out of all proportion to their military significance. During the Spanish Civil War the Fascist bombing of Barcelona in 1938 killed 1,300 civilians and wounded 2,000 more. From this British experts estimated that each ton of bombs dropped would inflict 72 casualties. This figure was then treated as definitive and used by the Home Office to predict a million casualties in London in the first few days of the Second World War. But this was alarmist nonsense with no basis in fact. The true statistics for Barcelona indicated a casualty rate of just 3.5 casualties to every ton of bombs – and that in an undefended city. In the entire Spanish Civil War just three per cent of the 500,000 dead were killed by air raids. Why were these statistics not published to allay public fears? The reason, clearly, is that they gave the lie to the claims of the air chiefs who had a vested interest in seeing that Bomber Command got the lion's share of the defence budget. The truth was that neither in 1938 nor in 1939 did Germany have the capability or the intention to launch a massive bombing campaign against London or other British cities. It is a sign of wholly inefficient intelligence work that the RAF was not aware of the range and potential of the main German bombers of the period. It seems that the RAF experts were more content to take guidance from science-fiction novels by H.G. Wells or from Alexander Korda's film *Things to Come*.

Ironically, when war broke out Bomber Command was itself quite unprepared for a major bombing campaign against Germany. Its pilots were not trained to penetrate enemy territory by day and find a target, let alone to operate at night. The British bombers themselves were mainly obsolescent – a remarkable situation for a nation that had virtually invented the notion that the bomber was the war-winning weapon. Britain had no dive-bomber, no ground support aircraft of any quality and no long-range bombers more recent than the feeble Whitleys and Hampdens. Thinking at Bomber Command seemed geared to an earlier, more gentlemanly form of warfare. It was geared to precision leaflet distribution, rather than for the terror raids that later became its hallmark. Inflicting casualties on the enemy was not Bomber Command's first concern. Bombardiers were even told to be careful not to drop a packet of leaflets all in one go lest they fell like a brick and injured somebody on the ground. Arthur Harris considered the leaflet drops as only useful 'to supply the Continent's requirements of toilet paper for the five long years of the war'. Hamburg and Bremen each received a million leaflets and the Rühr 3.25 million more. When it was suggested that the Black Forest should be ignited by incendiary bombs, Kingsley Wood, the Secretary of State for Air, replied in alarm, 'Are you aware that it is private property? Why, you'll be asking me to bomb Essen next.' While Wood was playing with toy soldiers, the Luftwaffe was razing Rotterdam to the ground with devastating terror attacks. The Dutch Foreign Ministry claimed that German terror bombing had killed 30,000 civilians. It looked as if Trenchard and Douhet were right. London again prepared for a million casualties in its first few days of bombing. But they were all wrong. The Dutch had panicked and exaggerated their losses. Just 980 had died in Rotterdam, yet the legend of that city's martyrdom fuelled the theory behind British and American saturation bombing throughout the war.

In fact, bombing in the early months of the war was highly inaccurate. German planes bombed neutral Dublin by mistake, and on another occasion bombed Freiburg, just inside their own border, instead of the French town of Dijon, over two hundred miles away. Navigation was quite appalling. Early British raids on Germany were every bit as bad. On 4 September 1939 an RAF attack on a German seaplane base resulted in 24 out of the 28 bombers involved being shot down, while the survivors managed to drop a few bombs by mistake on the Danish town of Esjberg – 110 miles from the target. Raids on German warships in Wilhelmshaven were just as disappointing. The few bombs that hit their targets either did not explode or bounced off the German armour, scarcely even scratching the paint. In addition, British casualties were heavy. Clearly the bomber did *not* always get through. Britain's response was to abandon daytime bombing in favour of night attacks. Thus began the concerted bombing campaign against Germany. Early raids on the Rühr produced 'significant successes'. However, photographic evidence did not support the claims of the bomber pilots. Two raids on oil installations at Gelsenkirchen by over 300 British bombers, dropping nearly 280 tons of bombs, achieved no hits and no damage to the plants. The truth dawned: 'the chances of a direct hit on a target were so small as to be negligible'.

As the evidence began to mount that precision bombing was a thing of the imagination rather than of reality, a shift in thinking took place. The British naval blockade in the First World War was reputed to have cracked civilian morale in Germany; it had become an article of faith among the Chiefs of Staff that German morale was more brittle than British morale and that the bombing of German cities might achieve a victory in the war without the need to commit massed armies of the kind that suffered on the Western Front in 1914–18. Thus the foundations of the Strategic Bombing Campaign were laid, based on the dangerous assumption that German civilians would be more likely to crack under the pressure of sustained bombing than their British counterparts. Intellectual arrogance and a peculiarly British chauvinism produced the following: 'The morale of the average German citizen will weaken quicker than that of a population such as our own as a consequence of direct attack. The Germans have been undernourished and subjected to a permanent strain equivalent to that of war conditions during almost the whole period of Hitler's regime, and for this reason also will be liable to crack before a nation of greater stamina.' This was a thin cement with which to build an entire aerial strategy. Hitler, for one, did not share the British belief that bombing weakened civilian morale. As the events of the London Blitz of 1940 had shown, the suffering that the civilian population shared actually brought them closer together and stiffened their resolve.

The devastating effects of the Blitz on Sheffield High Street, 12 December 1940. In spite of the serious damage and heavy casualties they inflicted, such terror attacks failed to dent British civilian morale or reduce war production.

Between March and July 1940 Bomber Command attempted to eliminate the German battlecruisers *Scharnhorst* and *Gneisenau*, stationed at Brest. Even though they were attacked by a total of 1,723 sorties, in which 1,962 tons of bombs were dropped, the ships emerged largely unscathed, having been hit just nine times. Even worse was the photographic evidence of the bombing raids on the Rühr. They revealed that just one bomber in ten got within five miles of its target and that the bombs dropped by these planes had been sprayed over an area of 75 square miles around the target.

In simple terms, Bomber Command had been wasting the country's money and precious resources on futile – if prestigious – raids into Germany. They had developed neither suitable tactics nor equipment for night bombing. Navigational aids were deficient and the bomb sight used was highly inaccurate. The heights of absurdity were reached during the raids of 1940 and 1941, which killed more British aircrew than German civilians – hardly a record likely to crack even the 'brittle' German morale.

But this stark evidence did not convince men like Trenchard and Harris that their belief in strategic bombing might be misplaced. In May 1941 Trenchard still insisted to Churchill that absolute priority should be given to long-range bombing, as the one certain way of winning the war – without the need for Britain to fight the Germans on the ground. Fighters, ground support aircraft, photographic reconnaissance planes, coastal command aircraft and so on should all be sacrificed to the insatiable appetites of Bomber Command. This was surely mistaken, as the outstanding success of British fighters like the Spitfire and Hurricane against German bombers should have demonstrated to anyone with a mind less than totally closed to conflicting evidence. Trenchard was overlooking the lessons of the Battle of Britain. He was prepared to reduce Britain's defensive air capacity to such an extent that in the event of a future German bomber offensive, Britain would be helpless to resist. But Trenchard found a supporter in Sir Charles Portal, one of the senior strategists at Bomber Command, who continued to believe that German morale could be broken by heavy bombing. Thus a policy of terrorizing the German population became a fundamental component of strategic bombing in the collective mind of Bomber Command. In the eyes of the air chiefs the army's role was subservient to that of the heavy bomber; they would simply mop up after the bombers had passed. This was the same false argument as had been used to justify the artillery bombardment on the Somme in 1916 (see p. 153). American General H.H. Arnold was just as convinced as his British colleagues that Germany could be beaten by bombing alone, but Churchill was not so sure. He had seen the light at last and – good historian that he was – he knew that indirect warfare could only take one so far and no further. Eventually the main enemy army would have to be confronted and beaten if victory was to be achieved. He told the air chiefs, 'Even if all the towns of Germany were rendered largely uninhabitable it does not follow that the military control would be weakened or even that war industry could not be carried on.' Yet this did not prevent Bomber Command from concentrating on night-time saturation bombing, even though there was clear evidence that 50 per cent of all bombs were falling in open country, far from the targets. Since their bombers lacked the accuracy to hit specific targets, Bomber Command apparently decided that a policy of killing Germans – any Germans – and damaging civilian property would suffice.

The entry of the United States into the war allowed the Allies to maintain their bombing of Germany day and night without respite, by the end of the war killing 300,000 civilians and wounding 780,000. Such a massive aerial bombardment might be expected to have achieved the destruction of morale that the air chiefs had predicted. Instead, as Churchill had warned, three years of bombing seemed to have stiffened German morale, as had happened in Britain during the Blitz. A further point, never mentioned by supporters of strategic bombing, is that two-thirds of all Germans never suffered any bombing at all. How their morale was supposed to be affected by bombing was never considered by the British and American air chiefs.

The most devastating indictment of strategic bombing was the fact that, in spite of laying many German cities to waste and reducing Hamburg and Dresden to ashes, it failed to prevent production of war materials in Germany. In fact production actually increased up to the end of 1944. In January 1945 a total of 30 new U-boats were built – a record for the entire war – even though German shipbuilding yards had been subjected to saturation bombing. Trenchard and Douhet were seduced by the myth of the effectiveness of saturation bombing because they had one-track minds. Churchill was seduced at first, understandably so, because he was desperate to save a new generation of British youth from the horrors of Passchendaele and the Somme. Harris and Portal were seduced because they believed that the Germans could never stand what the British had stood in 1940. Even after having reduced German cities to rubble 'Bomber' Harris missed the point that the rubble was still German rubble. It was not until their armies were beaten, their country invaded and their cities occupied by foreign armies that the Germans finally succumbed.

'The aeroplane is not a defence against the aeroplane.'

Hugh Trenchard in 1916

The failure of Trenchard – a leading exponent of the view that 'the bomber would always get through' – to appreciate the potential of the fighter plane meant that most of the funding for aircraft construction and development went to Bomber Command. As a result the RAF was deprived of top-quality fighters until the late 1930s, almost too late to prepare an effective defence against the Luftwaffe.

The phantom air force

The Italians took to the air with all the panache of true showmen. This new element gave a fresh impetus to their desire for triumphs of the soul. Let the Germans exult in their conquest of the earth like the burrowing dwarves of their *Nibelungenlied*, it would be Italians like Gabriel D'Annunzio, Giulio Douhet and Italo Balbo who would set the pace up among the clouds. Mussolini himself was caught by the flying bug and by the Italian thirst for superlatives – 'higher', 'faster', 'further'. In the early 1930s, while other nations were developing the military potential of aircraft, Italians were achieving aerial propaganda triumphs: breaking speed records, making formation flights across the Atlantic. It was Benito Mussolini's Flying Circus – virtual reality, but not reality itself. The Italians were not masters of the air; they had become its clowns.

Fascism had created the Italian air force, but it had created it in its own image. From the start appearance was everything and substance nothing. It was enough for a plane to look good, for an engine to sound good or a pilot to be smartly dressed. What happened when they met the foe was in the lap of the gods. And surely the God of War would be on Italy's side when the time came. An early influence on Italian air policy – as indeed he was on that of other nations – was General Giulio Douhet, who predicted that future wars would be won by bombing alone. Curiously, the state whose bombers everyone feared the most – Germany – was less influenced by Douhet's predictions than any other. Mussolini, however, was impressed by Douhet's apocalyptic scenario of bombers raining fire, disease, poison and lethal gases from the sky onto a defenceless civilian population. It all sounded so easy – as easy, in fact, as dropping mustard gas on helpless Ethiopian tribesmen – and even Italians could do that. Douhet was convinced that densely populated England would be particularly vulnerable to bombing. If Italy built the right bombers London could be destroyed within days. But Douhet was talking dangerous nonsense. He failed to appreciate that no weapon in history has ever reigned supreme for long without producing its own antidote or defence. Anti-aircraft guns, fighter planes and radar were developing at a faster rate than bombers, and were achieving such a level of effectiveness that unprotected flights of bombers were inviting certain destruction each time they took to the air. But Mussolini was not interested in counter arguments. Douhet's ideas had caught his imagination and all other considerations fell into the category of petty detail. Mussolini was particularly excited by Douhet's suggestion that the airborne bombers could be protected by floating smoke screens, from whose cloak of invisibility they would pour forth death and destruction. Mussolini was enthralled by the idea of his bombers dropping enough poison gas to kill five million Londoners in the first week of war.

Yet if Douhet was the prophet of Italian air power, Italo Balbo and not Benito Mussolini was his true disciple. Balbo was a far more dynamic character than Il Duce and he had an international reputation as a courageous and skilled pilot. Mussolini both admired and feared him at the same time. He once said that Balbo was the only Italian who might one day kill him. The problem for Mussolini was that Balbo, however high he flew, still managed to keep his feet firmly on the ground. When Mussolini boasted that he would black out the sun with the numbers of his planes and that Italy's air force was 'second to none', Balbo had that awkward way of telling him that this was nonsense. Balbo pointed out that his flatterers had been juggling the figures to mislead Il Duce and that Italy's real strength was about on the level of one of the Balkan states. Mussolini did not like hearing this and, on one occasion, even dismissed the editor of a newspaper who printed a picture of Balbo instead of himself. In the end Mussolini found an assignment for Balbo in Libya, far away from Rome. Mussolini had by now taken over the Air Ministry himself and did not want Balbo telling him what a mess he was making of it. Il Duce liked to tell the public that he spent the greater part of every day supervising each aspect of Italy's aircraft production. However, a senior civil servant at the ministry later reported that he had never once seen Mussolini in the Air Ministry building. Mussolini was the ghost in a machine that was itself a phantom. He believed that if he could 'wish' a bomber all the way to London, and boast about it enough, then it would happen. He once announced that he had had a plane developed that could destroy the entire British fleet in a single day, and one that could fly to England and back without being heard. He must have been deaf to the sniggers of his own colleagues.

And so the phantom airforce continued to grow. In 1935 an aeronautical experimental station was set up at Guidonia which contained one of the first wind tunnels built in Europe. Everything pointed to Italy being in the forefront of aeronautical research. It was a triumph for Fascism, but in every other way it was a 'white elephant'. At a time when vital technological developments such as automatic flying equipment, gyroscopes, high octane petrol, anti-icing devices, variable pitch propellers and retractable undercarriages were being made in Britain, France, Germany and the United States, none of these things was made at Guidonia. The fine buildings were the home of VIPs and foreign delegations, crowded around with pressmen and photographers. It was all part of the show.

When war broke out in 1939 Mussolini began to flex his military muscles by announcing that Italy had 8,530 aeroplanes. It was an awesome figure – but it was pure fantasy. Privately the Air Ministry reported that the correct figure was 3,000 front-line planes. Unfortunately, civil servants could not even tell each other the truth. The figure was next reduced to 1,000. Dawn was breaking and the phantom air force was withering away. The next figure was just 454 bombers and 129 fighters, all inferior in speed and equipment to existing British machines. When Mussolini heard the truth he refused to believe it and ordered its absolute suppression. Stories began to trickle out about the planes being flown from one airfield to another so that they could be counted over and over again and so boost the figures. Mussolini was desperate to stop the Germans finding out. He told them that Italy was producing 500 planes a month – the true figure was just 150. What is truly incredible is the fact that Italy built more planes during the First World War than during the Second. As Air Minister Mussolini cannot escape the blame for this disgraceful state of affairs.

A further negative result of Mussolini's aerial follies was that he rejected calls to

Wrecked Italian fighters at Catania, Sicily, 1943. The performance of Italy's air force in the Second World War belied Mussolini's claim to have enough planes to wipe out London within a week.

build aircraft carriers on the grounds that Italian land-based bombers could reach any part of the Mediterranean. The latter was completely untrue and the former a strategic *gaffe* of no small proportions. Had Italy possessed aircraft carriers instead of her new battleships it might have changed the outcome of the fighting in the Mediterranean. British positions at Gibraltar, Malta and Alexandria could well have proved untenable, and strong air support for Italo-German forces under Rommel might have reversed the outcome at El Alamein in 1942.

Mussolini may have been deceived by others, and may even have deceived himself, but in 1940 he knew that Italy did not have an adequate air force to fight a war against major powers like Britain and France. In North Africa Italian planes suffered from inadequate sand filters which led to the majority of them being grounded by engines clogged with sand – an inexcusable state of affairs in view of Italy's long experience of war in desert regions. Italian bombers – the Savoia-Marchetti S.79s – were apparently so dangerous to fly that no Germans dared fly in one. But towering above all other problems was the lack of a watercooled engine, something the British and Germans had developed years before. In 1934 the Italians had decided in favour of radial engines, light, simple to maintain, and strong enough to power the fabric biplanes that were then used to overawe African tribesmen. But once the British developed Hurricanes and Spitfires and the Germans Bf109s, the Italian fighters were left light years behind. Their radial engines were too feeble to drive modern planes equipped with armour, heavy machine guns and self-sealing fuel tanks at 400 mph. But the Italians had made their choice: clearly they saw their future in laying waste unprotected Ethiopian villages, not in taking on the modern technology of major military powers. They had boasted of their capacity to destroy Addis Ababa, Gondar and Harrar and to burn the entire Somali bush. But Douhet had boasted of destroying London! Up until 1940 aerial warfare had been easy and the *Regia Aeronautica* had not developed tactics for night flying or operating in bad weather outside the Mediterranean theatre of war. Their officers were dilettantes, who had joined the air force because they liked the uniform or enjoyed boasting to their girlfriends. When war came they had to be sacked in droves for incompetence.

For an air force built to match the boastings of Giulio Douhet the Italians had curiously spurned the heavy bomber in favour of light and medium ones. This influenced the size of bomb that could be carried and may explain the Italian preference for very small bombs, particularly 50- and 100-pounders. Italian pilots preferred to bomb from extreme altitude – this made hitting the target very difficult, but ensured that the ground defences could not get at them. A further problem with the light bomb was that even when a direct hit was achieved – and these were few and far between – little damage was done.

When, in late 1942, a four-engined bomber was designed – the Piaggio P108 – it was immediately nicknamed 'the flying weakness' because it was so mechanically unreliable. Pilots hated it bitterly since it had a habit of pitching them into the Mediterranean. An Italian-designed dive-bomber was briefly available, but was more dangerous to its crew than it was to the enemy. In the end the Italians were forced to buy the Ju87 Stuka, just at the point when the Germans had stopped using it as it had become 'easy meat' for the new Allied fighters.

The Italians also had trouble with their bombs. The fuses were often unreliable, the casings broke on contact with the target and incendiaries sometimes burst on launching, incinerating the aircraft. All in all it was a dangerous job dropping Italian bombs. When two pilots showed supreme initiative by disguising their planes as British Hurricane fighters and flying over the aircraft carrier *Victorious* as if about to land, the two bombs they dropped – which scored direct hits on the carrier's flight deck – both failed to explode.

Italian bombers also lacked an intercom between pilot and crew. If the bombardier wished to communicate with the pilot – to ask for a second run, for example – he had to leave his position and crawl into the cockpit to talk to him there. Lack of navigational aids led to a lot of 'blind' flying. When, in September 1940, 77 Italian bombers were flown to Belgium to help the Luftwaffe in the Battle of Britain, five got lost on the way and crashed, while 14 others succumbed to fog, rain and icing. In the war in the Mediterranean and North Africa the Italians felt more at home. In Libya the Italian transport planes did sterling service providing their troops with the necessities of war, 'from torpedoes to women' and 'from cannon to board games'. The Italian ground crews in the desert had an unusual approach to aircraft maintenance. Slipshod it may have been, but it was certainly original. Deprived of essential supplies from Italy the Italian mechanics relied on barter, making 'unheard-of deals . . . with mysterious Arab traders' who preyed on wrecked British and Italian aircraft. Mussolini's cheeks would have turned red if he had ever heard how the pride and joy of Italian Fascism, the *Regia Aeronautica*, was kept in the air by Arab scrap metal dealers.

To add to the design faults of the planes and their engines, and the poor quality of its officers and air chiefs, the Italian air force also suffered the shortages inevitable in an economy not geared to war. During the period 1937–43 the Italian aircraft industry actually exported four billion lire's worth of machines, engines and aircraft accessories at a time when Mussolini claimed his country was preparing for war. With supplies of desperately needed raw materials unobtainable, Italy entered the war against Britain and France in 1940 with enough aircraft fuel to last six weeks. Even this was subject to wastage because of the Italians' preference for tin-lined storage tanks rather than steel ones. The air force's lifeblood literally soaked away into the earth or the desert sand.

For a nation that took its football seriously Italy's air war began with a spate of alarming own goals. A naval engagement off Calabria between the Italian and British fleets was interrupted by the *Regia Aeronautica* which, unable to differentiate friend from foe, blithely bombed both sides. Fortunately, so inadequate had been the training of pilots and bombardiers, that there was never any real chance of them hitting anybody. More serious, though perhaps not to Mussolini himself, was the elimination of his rival, Marshal Italo Balbo, in a 'friendly fire' incident. Balbo had been inspecting the Egyptian front and was flying back to his base at Tobruk in a Savoia-Marchetti S79 bomber. The shape of the plane was quite distinctive but this did not save him. Panicking at rumours of an RAF mass attack, the Tobruk anti-aircraft gunners shot him out of the sky. With Balbo dead the slender thread that held Italian forces in North Africa together finally broke. Moreover, Balbo had been the only man who had the courage to remind Mussolini that all his boasting and wild assertions were mere bluff. By 1940 the bluff had gone too far and the Italian people were going to suffer the consequences of Il Duce's folly. And their phantom air force would be helpless to save them.

'When you have a fine plate of pasta guaranteed for life, and a little music, you don't need anything more.'

Italian General Ubaldo Soddu on his military philosophy, 1940

The Italian Way of War. This remarkable admission by a leading Italian commander characterized the Italian war effort in Greece and North Africa in 1940. While Soddu was supposed to be commanding Italian troops in the Greek theatre, he spent most of his time composing film music.

The general purpose dud

It has been shown that half the bomber sorties made over Germany by RAF Bomber Command in the Second World War were a complete waste of time. The weapon with which British bombers were armed – the general purpose bomb – was so unreliable that it is not stretching the truth to say that half the lives lost by Bomber Command in their night raids over Germany were unnecessary, and that the bombardiers might as usefully have been dropping leaflets as they did at the start of the 'Phoney War.' The Official Historian of the Strategic Air Offensive, Sir Charles Webster, was scathing about the general purpose bomb:

> Between 1939 and 1945 Bomber Command dropped over half a million 500lb G.P. bombs and nearly 150,000 two-hundred-and-fifty-pounders. Not only were these bombs often unsuited to the task for which they were used because of their general characteristics, which consisted of an unhappy compromise between strength of casing and weight of explosive, but they were also relatively inefficient and all too often defective weapons.

The fact was that the British Air Staff was antiquated in its thinking; as late as 1935 it had argued than nothing larger than a 500-lb bomb would ever be needed. Even when they did wake up to the need for bigger bombs in 1940, their 4,000-pounder gave a feeble blast for its weight and usually broke upon impact without exploding. It was estimated that nearly 40 per cent of all bombs dropped in 1940 failed to detonate. The development of the huge 'blockbusters' should have improved matters but did not, since early versions were dropped by parachute and were almost impossible to aim. It was an exasperating state of affairs. After the evacuation of British troops from Dunkirk, bombing was the only way in which Britain could strike back at Germany – yet Bomber Command could not even produce a worthwhile bomb to use against the enemy. Thousands of brave men lost their lives carrying defective bombs to Germany. It was criminal negligence – as bad as sending men into battle with their rifles loaded with blanks.

'I can't write what I mean, I can't say what I mean, but I expect you to know what I mean.'

Marshal of the RAF Lord Trenchard

Great minds in aviation could be just as difficult to work with as those in the other services.

Professor Willy makes a mess of it

It is a moot point as to whether the sailor or the airman is more at the mercy of the machine that bears him into battle. In view of the fact that men can swim, but cannot fly, one is forced to plump for the pilot and his aircrew. But as if aerial warfare was not dangerous enough in itself, pilots have sometimes found themselves obliged to take to the air in devices that hardly deserve the name of 'aircraft' at all. No one country has had a monopoly in producing 'aerial dodos'. Even such a renowned aircraft builder as Professor Willy Messerschmitt had his 'off days', and Germany suffered as a result. His worst moment came with his design for the Me-210, a twin-engined fighter, capable of doubling as a ground-support bomber. The Me-210 had been the brainchild of Messerschmitt's chief designer, Waldemar Voigt, but unfortunately Messerschmitt could not refrain from tampering with Voigt's blueprints to lessen the plane's weight and increase its wind resistance. The result was a wholly different plane, but this did not stop the Luftwaffe from ordering a thousand of them, even before they had seen the prototype fly. Such was the urgency of the military situation that Ernst Udet – then Director of Air Armaments – cut corners, with disastrous consequences. When test models of the Me-210 were flown, they went into a flat spin and their undercarriages collapsed on landing (Messerschmitt had weakened the landing wheels from Voigt's design to lighten the plane). From November 1941 special components and sections for the Me-210 arrived at the Messerschmitt factory from firms all over Germany, but not one acceptable plane was produced. Not only was the saga of the Me-210 bankrupting the company, it was also seriously damaging Germany's war effort. When enough planes were eventually produced to equip a squadron, the result was a catastrophe, with planes crashing and pilots killed. The squadron was promptly re-equipped with a different plane after 17 lives had been lost in a single week. With the Luftwaffe short of planes, and the Ju87 Stuka now vulnerable to improved Allied fighters, the need for the Me-210 was desperate. When Goering visited the Messerschmitt factory at Augsburg he found Messerschmitt 'a broken man. He was physically at a very low ebb and crazy with emotion. He was crying like a baby.' More than 370 half-finished Me-210s littered the factory, with components for hundreds more piled up. But still the plane would not fly.

Messerschmitt's tampering with Voigt's design had ruined his company and cost Germany a thousand planes when she most needed them. Eventually, in September 1942, the ill-fated Me-210 was redesigned from Voigt's original plans and tested with new engines. It was a success and was renamed the Me-410. Messerschmitt – whom Hitler described as having the skull of a genius – had caused his country irreparable damage. So frustrated had Goering become that he suggested an epitaph for himself: 'He would have lived longer but for the Me-210.'

'That will never make a fighter.'

Ernst Udet on first seeing the Bf109 in 1934

A classic misjudgment by the head of the Luftwaffe. The Messerschmitt Bf109 became one of the foremost fighter planes of the Second World War.

Willy's 'white elephant'

War makes men do strange things. It may make the steadiest of men reckless or the most carefree and confident dogged by self-doubt. And working for a man like Adolf Hitler can hardly have helped. Hitler knew the aircraft he wanted and regardless of its technical viability he insisted that he should have it. Satisfying such demands was clearly too much for Professor Willy Messerschmitt and prompted him to act like the fool he certainly was not. Gigantism – a product of megalomaniac vision – was never far from the surface of Nazi building, both in architecture and technology, and it reared its undoubtedly ugly head when Hitler decided that he needed a plane to carry tanks and other heavy equipment across the English Channel. In the autumn of 1940 'Operation Sealion' – the invasion of England – was still on the agenda and the head of the Luftwaffe, Hermann Goering, was responsible for ensuring that German troops reached England in one piece, and with enough equipment to fight. He therefore turned to Messerschmitt to supply the answer.

At first Messerschmitt thought along the lines of a huge glider, towed by four Ju-52 transports. In March 1941 the glider was ready. Nicknamed *Gigant*, it was a vast structure – bigger than any existing bomber or transport – with a wingspan of some 60 yards, and a cargo bay capable of carrying a light tank. To all appearances Messerschmitt had produced a winner. Unfortunately, towing such a monster was a real problem. An airliner failed to manage it, as did three Me-110 fighters, and success was only achieved by bolting together two separate He-111 bombers. But by the time Messerschmitt had worked out how to tow the *Gigant* 'Operation Sealion' had been cancelled and the glider was no longer needed.

But Messerschmitt did not give up. He and his engineers now produced a *Gigant* with engines – the Me-323. The problem was that the new plane was so slow that it seemed to be stationary in the sky. The *Gigant* had less than half the speed of the British and American fighters it might meet. With this in mind one can only wonder at the madness of sending a squadron of the Me-323s to the Mediterranean front where they were bound to meet the fighter aircraft that would prove their nemesis. On 22 April 1943, Rommel's army in Tunisia was in desperate need of supplies of all kinds and 16 *Gigants* were loaded with over 300 tons of food and munitions. Over the Mediterranean these 'sitting ducks' encountered RAF fighters which proceeded to massacre them, only two of the 16 Me-323s surviving to reach Tunis. The fate of the *Gigant* had been a foregone conclusion and the remaining planes were transferred to the Eastern Front where the Luftwaffe still enjoyed aerial superiority

The Messerschmitt Me-323, popularly known as the *Gigant*, proved a costly failure for Germany during the Second World War. Despite its impressive freight capacity it was painfully slow and succumbed easily to Allied fighters.

and was able to escort them on all their flights. By the time production of the Me-323s was closed down in March 1944, 201 of them had been built. But as the Soviet fighters gained the ascendancy in the east there was nowhere for these monster planes to hide. They had used up a large proportion of Germany's aerial production capacity to no useful purpose. Their lack of speed made them helpless against modern fighter aircraft. One is forced to wonder whether the technical interest of producing the *Gigant* caused Messerschmitt to overlook the fact that it was in reality a 'white elephant'.

The Buffalo – an endangered species

The outbreak of war in 1939 was a testing time for the military hardware of the combatant nations. Ships, planes, tanks or guns now faced the only test that mattered: how would they stand up against the best the enemy could throw at them? Naturally there were many success stories, but there were just as many failures and some of these were so severe that they had a serious effect on the outcome of the fighting. One such failure was the American-built Brewster Buffalo fighter, a combat aircraft of quite appalling inadequacy. The Buffalo was heavy, sluggish, poorly armed and fatally underpowered. Only Britain's desperate need for fighters of any kind could have persuaded the RAF to purchase nearly 150 of them. Compared to the latest British fighter – the sleek Spitfire – the Buffalo was woefully clumsy. And the idea of the

The chunky, American-built Brewster Buffalo fighter was used by the RAF in the Far East, but found itself totally outclassed by the Japanese Mitsubishi Zero.

Buffalo contesting the skies with the rapier-like Messerschmitt Bf-109 was laughable. As a result the Buffaloes, nicknamed the 'peanut specials', were consigned to Britain's bases in Singapore, Malaya and Burma, on the assumption that anything – even the Buffalo – would be good enough to face the Japanese. In December 1941 – just days before Singapore was attacked – Air Chief Marshal Sir Robert Brooke-Popham had said publicly that, 'We can get on all right with the Buffaloes here. They are quite good enough for Malaya.' This decision – one of many that reflected a contempt for Japan as a military power – meant that the Buffalo would be pitted against the Mitsubishi Zero, one of the best fighter planes ever designed. The Vice-Chief of the Air Staff in Singapore reported that the Buffaloes 'would be more than a match for the Japanese aircraft, which were not of the latest type'. He was wrong, of course, and when the Buffaloes did meet the Zeros they were soon as extinct as their animal counterparts on the American plains. On American Pacific airfields Buffalo pilots knew they had a one-way ticket if they were expected to encounter Zeros. By the summer of 1942 no more Buffaloes were sent into action.

Lessons from the 'Pulpit'

The design of the Royal Aircraft Factory BE.9 – nicknamed the 'Pulpit' – would be a strong contender in any competition to find the least 'user-friendly' aeroplane ever conceived. Its principal defect – no doubt seen as its greatest strength by the designer when deep in his cups – was the decision to place the gunner in a 'pulpit' in the nose of the plane, in front of the propeller. There was no shield between the gunner and the propeller, thus as the gunner struggled to control his Lewis gun, swinging round to fire at passing Huns, he was literally only inches away from the gyrating blades. Thousands of feet up in the air he would find that the wind, combined with the suction of the propeller, threatened at any minute to drag him backwards to a grisly death. How anyone could keep his mind on his job with the whirling blades of the propeller in such close proximity is difficult to imagine. In addition, since the gunner and the pilot were separated by the propeller, they had no way of communicating with each other. It comes as no surprise to learn that the Pulpit was an unpopular plane that saw little service and was scrapped in 1915.

Lack of vision

Not content with the appalling design of the BE.9, the Royal Aircraft Factory surpassed themselves with the ridiculous RE.8, a two-seater reconnaissance plane that presented more than its fair share of problems for the unwary pilot. Powered by a feeble 150 hp engine, the RE.8 had trouble reaching altitude – taking some 45 minutes to reach 10,000 feet – and once up there had extreme difficulty in maintaining it. Its inadequate tail made it prone to uncontrollable spins, while the placement of the engine made it almost impossible for the pilot to see over the top of it, so that every landing was more or less 'blind'. To make matters worse, the fuel tank was placed so close to the engine that fires were frequent. The plane was very difficult to balance and pilots were warned not to take off without at least 150 pounds in the gunner's compartment – only well-rounded gunners were considered for this aircraft.

Many American companies began building military aircraft after the First World War. One of these was owned by Grover Loenig and in 1921 he was commissioned to build new fighters for the US Army. The result was a high-winged monoplane, known as the PW-2. The problem was that Loenig seemed to have completely ignored the question of visibility. The wing of the PW-2 was attached to the fuselage in such a way that the pilot could see nothing above him (a serious problem for a fighter pilot) and virtually nothing in front of him (a problem for any pilot). The plane's pilots must have often wished they could see through the wing – indeed, one Lieutenant Harold Harris had his wish granted when the wing of his plane fell off during a flight in October 1922. Harris lived to tell the tale – which is more than the PW-2 design did – and became the first American to escape by parachute.

Four May days

During the campaign in France in 1940 the RAF employed an Advanced Air Striking Force (AASF) of ten squadrons of light bombers – Fairey Battles and Bristol Blenheims. The Fairey Battle was a first-generation monoplane bomber of archaic design and flight characteristics. Its deficiencies were clear even in 1936, when further orders for the plane were cancelled. But in 1940 the RAF was stuck with it. It was far too slow and awkward for the fighting conditions in France and, with a maximum speed of just 240 mph, it was an easy prey for both German fighters and mobile anti-aircraft guns. Aware of its vulnerability to fighter attack, British commanders ordered its pilots to make low-level runs at under 250 feet. But these proved disastrous in the face of the 20-mm Vierling anti-aircraft batteries that accompanied all German columns.

The early days of May 1940 proved too much for the Battle design. On 10 May, for example, 32 Battles attacked German columns in Luxembourg and thirteen were shot down by machine-gun and small-arms fire, while all the others were damaged, more or less seriously. The following day the Battles tried again; eight began the attack and only one returned. On 12 May, all five Battles that attacked strategic bridges were shot down by ground fire. These planes had been flown by volunteer crews – by this time it was obvious that to fly a Battle into action was virtually suicidal.

The Fairey Battle light bomber was already out of date by 1939. Its woeful performance in the battle for France in May 1940 starkly revealed its limitations.

One pilot who survived the inferno and parachuted down inside German lines was actually lectured by the German officer who captured him on the futility of such low-level attacks against heavily defended areas: 'You British are mad. We capture the bridge early Friday morning. You give us all Friday and Saturday to get our flak guns up in circles all round the bridge, and then on Sunday, when all is ready, you come along with three aircraft and try to blow the thing up.'

In just three days RAF strength in light bombers had slumped from 135 Battles and Blenheims to just 72, a rate of attrition almost without parallel in the history of aerial warfare. The Battle pilots were now told to change their tactics and to try to use their planes as dive-bombers – a role they were not designed for. On 14 May the AASF made a last desperate attempt to destroy the German bridgeheads on the Meuse. Out of 71 Battles and Blenheims that attacked, 40 were shot down by German fighters and ground fire. On this one day, 62 per cent of Battles sent out failed to return. This was the highest rate of attrition in the RAF's history and with each plane lost irreplaceable pilots and aircrew were needlessly sacrificed. In just four days 86 Fairey Battles had been destroyed. And all the time the Ju87 Stuka – as slow as the Battle but designed as a dive-bomber – was winning the battle for France. While the Battle continued its suicidal attacks, the lack of a real British dive-bomber was allowing German armoured columns a free ride across France.

'All our types and all our pilots have been vindicated as superior to what they have at present to face.'

Winston Churchill in 1940

This is a typical example of Churchillian bombast. However high the quality of British pilots was, the planes they flew were often inferior to German and Japanese equivalents. Britain's failure to face the truth of this contributed to disasters in France in May 1940 and Malaya in December 1941.

CHAPTER 4: SAWDUST CAESARS

The military profession has often been beset by interfering politicians. As 'disciples' of a higher calling – that of 'grand strategy' – these meddlers have underrated military science and have extolled the virtues of the mind uncluttered by the dogma of a soldier's education. In the 20th century the most active of these meddlers have been Adolf Hitler and Winston Churchill – both of them with some experience of war, but none whatsoever of the problems of military leadership. Yet each found himself uniquely placed to interfere with the decisions of generals and admirals. And because each felt that he knew better than the entire body of general officers, he felt confident to make decisions that affected the entire progress of the Second World War. Not far behind these military dilettantes came the original 'Sawdust Caesar', Benito Mussolini, who shared with Churchill and Hitler just enough military experience to make him dangerous. Believing, like Hitler, that war was a matter of willpower, Mussolini indulged in a series of ill-conceived military operations that he willed to succeed in spite of the logic that says that a man wearing cardboard shoes will soon have frozen feet in a Russian winter, and that soldiers using dummy weapons will soon succumb to those using real ones. Sawdust Caesars do not see things this way. Churchill had dreams in which he saw multitudes of Balkan tribesmen – Greeks, Slavs and even Turks – leaping like ravening wolves to rip open the soft underbelly of Germany. It did not happen, either in 1915 or 1940.

Winston Churchill had an ambivalent attitude towards generals. On the one hand his experiences in the Sudan under Kitchener and in the Boer War encouraged him to despise them for their weaknesses, yet as a descendant of the great John Churchill, Duke of Marlborough, he wanted above all to be a super-general himself. The result of his early exposure to military campaigning was that he grew up thinking that he knew better than the military blockheads who frequently gained command through seniority. And – in both world wars – Churchill found himself in positions to interfere in both military and naval planning.

Churchill and Greece, 1940

Churchill's 'Greek gamble' was the worst of all his interventions in British strategical thinking in World War Two.

Driven from France in June 1940, Britain found that she could no longer strike at Germany, except through bombing raids on German cities. As a result, the decision was taken to give priority to the war in the Middle East. At least there Britain could fight Hitler's ally, Italy. In any case, Britain was committed to holding Egypt, which had been an important British military base for 60 years or more. But in assuming that his desperate need for oil would force him to attack the Middle East, Britain was misreading Hitler's strategic intentions. In fact such a move was never in Hitler's mind, and his intervention in the Mediterranean area was a defensive response to British pressure on Libya and Greece.

British Mediterranean strategy in the early period of the war was quite unrealistic, and lacked a sound awareness of military realities. Churchill hoped to win allies in the region, particularly Turkey and the French in North Africa, but he never had a serious chance of achieving this. Weygand, the ruler of French North Africa, may not have been pro-German but he was loyal to the Vichy Regime and would not have allied with the British unless they could demonstrate that their power was far stronger than it was. And the Turks were militarily too lightweight to be of use as an ally of Britain. If they had entered the war they would have become a drain on Britain's military resources.

Italy's entry into the war in 1940 offered Britain a chance to win easy and morale-boosting victories in Africa. Although she could not hit directly at Germany, she could damage German prestige by defeating Italy in Libya and Ethiopia, and victory in North Africa might eventually allow British troops to open a door to Europe by an invasion of Italy. Nevertheless, it was most important that Britain did not attempt to overstretch her limited resources. Churchill needed to be told that however pressing the political arguments might be there were some decisions that would have to be reached on military criteria alone. Unfortunately, Churchill's meddling in the military affairs of Middle East command in 1940–1 contributed to three separate yet connected disasters: the Greek expedition, the fall of Crete, and Rommel's re-capture of Cyrenaica (the eastern part of Libya). Although there were failures at tactical level in each of the above cases, the biggest blunders were committed by the political strategists.

General Richard O'Connor's dramatic destruction of the Italian armies in Cyrenaica gave Britain a boost at a time when she needed it most. It also startled the Germans into sending Rommel to try to restore the Axis position in North Africa. But having made such a good start, Churchill allowed his butterfly mind to find something else to land on: his own personal hobbyhorse – a Balkan alliance against Germany. All the military gains of professional soldiers were about to be frittered away by the gambling instincts of an amateur strategist.

In April 1939 Britain had guaranteed support to the Greeks in the event of a German attack. However, after the invasion of Greece by the Italians in October 1940, the Greek leader, Metaxas, was not eager to accept military support for fear of provoking German intervention. In fact, the Greeks were at least holding their own against Mussolini's incompetent commanders in Albania.

A.J.P. Taylor has described the decision to intervene in Greece as 'taken on political and sentimental grounds' – Greek independence, Lord Byron and all that. Churchill was trying to inspire Turkey and Yugoslavia to stiffen their resistance against Germany and to put heart into all freedom-loving nations. However, such rhetoric availed the people of the Balkans little. There were times when Churchill seemed to be at odds with his own century. The primitive Balkan armies were no match for German professionals. Yet Churchill was determined to support Greece at the expense of North Africa. He told the Chiefs of Staff on 6 January 1941:

> It is quite clear to me that supporting Greece must have priority after the western flank of Egypt has been made secure. . . . We must so act as to make it certain that if the enemy enters Bulgaria, Turkey will come into the war. If Yugoslavia stands firm and is not molested, if the Greeks . . . maintain themselves in Albania, if Turkey becomes an active ally, the attitude of Russia may be affected favourably.

But if the securing of the western flank of Egypt had overall priority, it surely meant that until the Italians had been finally beaten in North Africa it would not be safe to plan other large-scale operations. Yet this is exactly what Churchill asked Wavell to do, by drawing troops away from the victorious advance through Cyrenaica. Churchill was making military decisions purely on political grounds – that of giving help to the Greeks – apparently strengthened by Foreign Secretary Anthony Eden's opinion that the Balkans could be held regardless of the situation in Africa. He was ignoring those awkward problems that got in the way of his grand design. Shifting an army from North Africa to Egypt posed all sorts of problems for his generals, 'such as weather, distance, the breakdown of vehicles, the lack of spare parts, the dribble of petrol from ill-designed cans, and the physical and moral exhaustion of men pressed beyond endurance'.

The Greek commander, Papagos, was not very pleased by the British eagerness to intervene in Greece. In his opinion it made no sense. He felt it would be far wiser for Britain to concentrate on completing the conquest of North Africa before attempting anything else. But once started, Churchill – like a steamroller running downhill – was difficult to stop. Even though O'Connor and the 7th Armoured Division were on the point of complete victory, the general was told to stop on the orders of the Cabinet. On 12 February, Churchill sent Wavell a long telegram praising the success of the troops, but stopping any further advance. Only minimum forces were to be left in Cyrenaica and the largest possible force was to be assembled for despatch to Greece. It was hoped that at least four divisions, including one armoured, would be made available at once. All O'Connor's air cover except one squadron was to be withdrawn. Against the advice of two of his senior staff officers, Wavell agreed to back the Greek operation.

Meanwhile, political factors had prompted the Greeks to change their mind and accept British aid. The death of President Metaxas and his replacement by the more pliable Koryzis, had enabled Churchill to persuade the Greeks that it was in their interests to accept military help. But Hitler was not prepared to see the Italians defeated in Greece – this might lead to a British threat to his weak flank while he invaded Russia. He therefore sent Luftwaffe squadrons to Sicily from where they were able to shake the British hold on the central Mediterranean. And while Rommel prepared to intervene in Libya, Hitler invaded Greece. British resources were already fully stretched in Libya and East Africa and yet Wavell, as Middle East commander, now had to face a new campaign. On 7 March, with Wavell's

Winston Churchill in Karsh's famous portrait. Churchill's military intervention in Greece was an ill-conceived venture that saw the British driven out of the Balkans in a matter of weeks, and nearly led to the loss of Egypt.

agreement and the support of the Chiefs of Staff, 50,000 British troops were landed in Greece. General Alanbrooke was in despair, seeing the Greek adventure as a huge strategic blunder. As he said, 'Why will politicians never learn the simple principle of concentration of force at the vital point and the avoidance of dispersal of effort?' These words were meant for Churchill.

Within a month the Germans had invaded the country and driven the British out, with the loss of 12,000 troops and all their equipment. It was a massive blow to British prestige. All the high hopes that had followed O'Connor's triumph in Cyrenaica were lost. Thanks to Churchill's intervention Britain was down again – if not actually out for the count. Meanwhile, matters had taken a very serious turn in Libya. The arrival of Rommel had transformed the situation and with minimal forces – just 50 tanks – he was able to overrun the inexperienced troops who had replaced O'Connor's veterans. The defence of the Western Desert had been left to a skeleton force of an under-equipped and inexperienced Australian division and the so-called 2nd Armoured Division, which had just one brigade of worn-out tanks, no more than half of which were serviceable at any one time. By 11 April the British had been driven out of the whole of Cyrenaica, except for Tobruk, which was surrounded.

Churchill's attempt to establish a British foothold in Europe was a serious blunder. The rewards, in terms of a Balkans Alliance against Germany, had not been worth the risks involved. Churchill should have known that Yugoslav, Turkish and Greek armies could never stand against a German army that had overthrown France in a matter of weeks. The reduction of British strength in Egypt acted as an open invitation to Hitler to gain a foothold for himself in North Africa. Just as the Germans had revitalized their Austrian allies in World War One into achieving a great victory at Caporetto, so the provision of an outstanding commander – Rommel – together with German troops, equipment and air power, might restore the morale of the Italians.

Churchill had always seen Egypt as the strategic centre of Britain's Mediterranean strategy. Yet by undertaking the Greek campaign he was allowing the Germans to replace the Italians as a real threat to 'the jugular vein of the British Imperial system – the Suez Canal'. There really had never been any chance of the British and Greek forces holding back the power of a German invasion. Churchill's shortsighted strategy simply offered four British divisions as a sacrifice. Even Britain's much-vaunted sea-power only managed to evacuate the main part of the British force and offer it temporary relief on Crete.

Fortress Singapore

The fall of 'Fortress Singapore' to the Japanese on 15 February 1942 was possibly the greatest blow that the British Empire ever suffered. The loss of prestige was enormous. The Empire 'on which the sun never set' seemed to have been irrevocably damaged. And the man who deserved most of the blame for the Singapore fiasco was the one who had most personified the supposed indomitability of the British Empire through the difficult years of the 1930s – Winston Churchill. The rhetoric had come home to roost and its wings were bare of feathers. A military catastrophe occurred because politicians had concealed the truth for so long that there was no time to restore a sense of balance.

British naval power was the bedrock on which the Empire had been built. But as other powers developed strong navies, it became impossible for Britain to protect all her colonial commitments. After the First World War the emergence of Japan as a naval power, and one that was increasingly hostile to Britain, meant that the Admiralty had to juggle its forces to maintain a powerful Home Fleet as well as honour its commitments to Australia and New Zealand in the Far East. In 1921 the great naval base at Singapore was established as a means of reassuring Britain's eastern colonies. But during the inter-war period financial cutbacks meant that Singapore had to stand increasingly as a symbol of power rather than a manifestation of its actuality. Britain promised Australia that a fleet could reach Singapore within 70 days of a Japanese attack. And so the 'myth of the fleet' was born. This was good propaganda, but was it based on anything other than wishful thinking? Gradually the time schedule changed – to 90 days, then to six months, and finally in May 1939 it was admitted that no guarantees were possible. The navy was no longer capable of defending the east. The task would have to be left to the army and the air force at Singapore. The Australians might have accepted this had the British really meant it, and demonstrated their intentions by placing a strong garrison in Malaya and supporting it with a powerful force of modern planes. Instead, there was more propaganda. It was said that the jungles of Malaya were impassable to enemy troops, that huge naval guns pointing out to sea would sink any Japanese transports, and that the feeble, short-sighted Japanese fighting man would be no match for British and Australian troops. Japanese planes would be helpless in the face of even the second-rate aircraft that the RAF stationed in Malaya. Churchill must have known that all this was untrue, and yet he used it to calm the fears of the civilian population of Malaya and Singapore and to satisfy the Australian government.

Churchill began to speak of 'Fortress Singapore' and stressed that the Japanese could not strike directly at Malaya without bringing the Americans into the war against them. He claimed that:

> Singapore is a fortress armed with five 15-inch guns and garrisoned by nearly 20,000 men. It could only be taken after a siege by an enemy of at least 50,000 men . . . as Singapore is as far away from Japan as Southampton is from New York the operation of moving a Japanese Army with all its troopships and maintaining it during a siege would be forlorn. Moreover, such a siege, which should last at least four or five months, would be liable to be interrupted, if at any time Britain chose to send a superior fleet to the scene.

But Britain no longer had the capacity to send a 'superior fleet' to Singapore.

The strength of the RAF in Malaya was risible compared with the potential power of the Japanese. RAF planners had estimated that the Malayan command needed 556 aircraft. Even this would have been thin in the face of the 700 planes that the Japanese could use in the coming campaign. However, it was positively generous compared

British troops surrender to the Japanese at Singapore, February 1942, perhaps the lowest point of Britain's fortunes in the Second World War. Contrary to popular belief, the British garrison outnumbered the Japanese quite heavily, but capitulated to Japanese air power and more progressive infantry tactics.

with what Churchill eventually considered adequate. In 1941 the RAF strength in Malaya consisted of just 141 second-rate aircraft, 41 of which were the hopelessly outdated Buffaloes (see p. 70) that merely provided target practice for the Mitsubishi Zeros. Infantry strength was just 33 brigades, many of which were half-trained Indian units, instead of the 48 considered the absolute minimum by the Chiefs of Staff. How had this been allowed to happen? After all, the threat of Japanese aggression in the east had been apparent for a decade after 1931.

The answer was that Churchill had earmarked all his best available equipment for his new ally, the Soviet Union, after June 1941, and Singapore was given a low priority. While from July to November 1941 Britain sent more than 200 Hurricane fighters to Russia, Singapore had to make do with Brewster Buffaloes. When it was suggested that more planes should be sent to Singapore to raise its aerial strength to at least the 336 first suggested by the Chiefs of Staff, Churchill was furious:

> I do not remember to have given my approval to these very large diversions of forces. On the contrary, if my minutes are collected they will be seen to have an opposite tendency. The political situation in the Far East does not seem to require, and the strength of our air force by no means warrants, the maintenance of such large forces in the Far East at this time.

Having squandered so many planes during the ill-fated Greek operation, and having sent more to Russia, Churchill is estimated to have wasted 600 first-rate fighters, even part of which number could have turned the tide against the Japanese. But Churchill preferred to fight the Japanese by gestures, like the dispatch of Force Z, the *Prince of Wales* and the *Repulse*, which – without adequate aerial cover – were both sunk by Japanese aircraft only days after their arrival at Singapore.

With the air force inadequate and the navy destroyed, the defence of Malaya and Singapore was left to the army, whose garrison of 90,000 men was made up of many half-trained and second-rate units. Some Australian troops had sailed for Malaya only two weeks after their men had enlisted. The problem was that everyone had taken it for granted that these troops would not be facing front rank Japanese soldiers so early in the war. The collapse of the other two services meant that the army had to make do with what it had, however inadequate that might be. British troops sent to stave off disaster in Singapore were just as poorly trained. The 18th Division, for example, had been trained in mechanized warfare and earmarked for the Middle East. Instead they found themselves in the front line in Malaya without a single tank in sight! The fact that they were sent to Singapore only days before the city surrendered was a strategic absurdity. They were not acclimatized and, fresh from their transports, were not fit to go straight into action. As a virtual sacrifice to the Japanese, they might as well have been shipped straight to POW camps.

But Churchill faced an agonizing situation. The Australians were rightly outraged at the prospect of Britain abandoning her responsibilities towards them. They could not accept a craven surrender of Singapore. Yet the more troops that Britain locked up in the fortress the more would be captured by the Japanese when Singapore fell. It was a 'no-win' situation. Churchill was paying for years of empty rhetoric and decades of Treasury cutbacks. All he could do was to demand that every soldier fought to the last – hardly reassuring advice. But it was his responsibility as war leader to do more than merely try to bluff the Japanese, while at the same time misleading his allies into believing that the naval base was indeed a fortress. His interference in the dispatch of adequate aircraft to the region was a disastrous mistake. The resulting surrender of the entire Singapore garrison of over 100,000 men was the greatest military capitulation in British history.

Mussolini's 'tossed salad'

A politician who lies to himself has problems – both psychological and military ones. By 1939 Mussolini knew that Italy was not strong enough to join Germany in waging war against Britain and France. Yet a year later, with France beaten and Britain on the brink of defeat, he was able to convince himself that his country had the strength to fight Britain in North Africa as well as to invade Greece. What caused Mussolini's change of mind was less the growth of Italy's military power or even the weakening of his opponents, than pique at Hitler's successes, and the fact that the Führer did what he liked without ever consulting his Axis partner. Mussolini was determined to show that what Germany could do, Italy could do at least as well. While Denmark, Norway, Holland and France fell to Hitler's panzer divisions, Mussolini had it in

Benito Mussolini, the original 'Sawdust Caesar', in a portrait of 1937. The Second World War revealed the hollowness of Mussolini's pretensions to military leadership. Italy's war record was disastrous, her forces being defeated everywhere until they were baled out by the Germans.

mind to invade Greece, without consulting the Führer. As Hitler had done to him in the past so he would do in return – present his ally with a *fait accompli*.

As the original 'Sawdust Caesar', Mussolini combined supreme ambition with very limited ability – always a dangerous combination. He was a military dilettante, who seemed to think that time spent in the trenches in 1915 was enough education for a new Napoleon. He was full of rhetoric: 'Words are a beautiful thing, but rifles, cannons, warships and aircraft are still more beautiful.' He was never able to overcome the journalist in him, confusing style with content, complaining if a 'present arms' was shoddily done or if an officer could not get his legs up high enough to perform the ludicrous goose-step (introduced in 1938), but failing completely to notice the medieval bronze cannon towed in military displays for lack of modern equipment, or the dummy tanks that stood in place of the real thing. In June 1940 a senior officer in Greece complained 'the Duce made the visit more as a journalist than as a commander: no word to his staff; no visits to subordinate commanders; not a single conference with the officers, but only a rapid review of the troops in formation, often carried out without even descending from his automobile'. And with this facile approach to modern war Mussolini compounded the problem by appointing himself head of each of the three armed services.

The truth was that the Italian army was no more ready for war against Greece than it was for inspection by anyone who knew his echelon from his epaulette. Mussolini – a rank amateur surrounded by spineless sycophants – was embarking on a military adventure without the slightest knowledge of what it involved. He allowed himself to be persuaded that the war would be nothing but a 'promenade' and that the enemy 'was devoid of all military qualities'. In only one way did he succeed: in infuriating Hitler. When he heard that Mussolini had invaded Greece the Führer was quite aware of the likely consequences: the Italians would be beaten , since 'the Greeks are no mean soldiers'; the entry of Greece into the war would give Britain air bases on the flank of a German invasion of Russia and, worst of all, the Germans would have to go in to rescue their allies, resulting in a quite unnecessary Balkan campaign for Germany.

In attacking Greece, Mussolini had deliberately chosen what he thought was an easy option, but even so he was soon out of his depth. Brass cannon and dummy tanks cannot be used in real battles. And whose idea had it been to demobilize 300,000 front-line troops and release 600,000 reservists to help with the harvest only weeks before invading Greece? Mussolini was beginning to think he had been badly advised. The Greeks were fighting instead of running away. General Visconti Prasca had assured him that the war would consist of 'a series of rounding-up' operations. The only 'rounding-up' that took place in the first few days was of Italian prisoners.

Mussolini had been told to expect dissident groups inside Greece to overthrow the government there and present his troops with a walkover. But it never happened. And when Prasca described preparations for the campaign as perfect, Mussolini was fool enough to believe him. He should have known – indeed perhaps he did know at some basement level of his consciousness – that the Italian army was a sham. It was no more modern than the Greek, and was manned by too many young reservists. It had been his decision, if he was honest, to send the experienced men to pick the grapes. The numbers of divisions had only been published to impress foreigners; he had increased their numbers by reducing each three-brigade division to two brigades or even just a single one. And many of these brigades were made up of blackshirts rejected from the army for reasons of physical – or mental – unsuitability. The old-fashioned Italian tanks were ineffective in the muddy conditions – although the cardboard ones were at least of a modern design. There was little anti-aircraft defence, and – although 1,750 lorries had been earmarked for Greece to begin the attack – just 107 turned up in time.

In the first two days of the invasion the Italians made little progress. Divisions and brigades were commanded by officers who had only just arrived in Greece and had not seen their commands before. And in the torrential rain everything bogged down in the mud. The transport system, both by air and sea, was run by rank incompetents. Supplies of all kinds were packed piecemeal, turning the task of finding what was needed into a sort of crazy jigsaw puzzle. Weapons parts, radio transmitters, field kitchens, blankets, baggage, medical supplies, ammunition, and animals were packed exactly as they arrived at the ports, and even in the same containers. The soldiers often found themselves being sent up the line to the front with just light arms, no ammunition, six pairs of socks and a copious supply of prophylactics. In the province of Apulia 30,000 transport animals waited with their drivers, but there was no way of getting them to Greece. Eventually, some of them broke free and rushed around the countryside causing chaos.

When the Under-Secretary for Air, Dervisciani Pricolo, asked the commander-in-chief, Visconti Prasca, what news he had of the Julia Division, he was alarmed to hear that nothing had been heard from it. He asked if that was not strange, as the invasion had begun five days before, but Prasca assured him that there was a perfectly simple explanation: the division did not have any radios yet and, anyway, the storms had blown all the aerials down.

The entire campaign was falling apart. One commentator called it Mussolini's 'tossed salad'. Amidst the sound and fury General Ubaldo Soddu, sent to replace Prasca as commander, soothed his artistic soul by composing film music. This was just the sort of thing that Mussolini had feared. How could he expect the Italians to make tough soldiers when their commanders behaved in this way? Mussolini felt obliged to dismiss Soddu 'for reasons of health'. In an attempt to set a good example, Il Duce ordered ministers and leading Fascist officials to go to the front in Greece as volunteers. There they would set an example like the Roman emperors of old; eating with the men, sleeping with the men, fighting with the men and dying with them if necessary. This was an unattractive prospect to most of his senior officials, who were used to parading around various divisional headquarters, contradicting the orders of the military commanders and generally getting in the way. Mussolini's idea led to many absurdities; in several cases totally unqualified bureaucrats were given vital front-line commands, and on one occasion official documents had to be flown out to a minister in Greece and signed by him in the heat of the battlefield. Another minister amused himself by bombing Corfu, because it had no anti-aircraft defences.

Another insisted on having an escort of fighter planes every time he flew anywhere. Most, however, kept as far from the fighting as possible. It was not the spirit that Mussolini hoped to instil in the Italian people. As severe winter weather set in and 13,000 Italian soldiers – kitted out with cardboard boots – developed frostbite, Mussolini exulted in the martyrdom of the Italian people: 'The snow and cold are fine,' he said, 'they will kill off the weaklings and improve this mediocre Italian race. One of the principal reasons why I wanted the reforestation of the Apennines was to make Italy colder and more snowy.'

While Mussolini played at war, the Greeks were taking it all very seriously. As one Italian soldier wrote home bitterly: 'At school they had heard that it was a fine thing to die with a bullet in one's heart kissed by the rays of the sun. No one had thought that one might fall the other way with one's face in the mud.' Mussolini's mad Italian adventure ended in tragic defeat. Only the intervention of Germany prevented the Italian armies being driven into the sea. Even for a 'Sawdust Caesar' it had been a pitiful display of military incompetence.

Forging an Iron Lady

For prime ministers and presidents a foreign policy crisis can come at the most inappropriate time; one has only to think of President Jimmy Carter and the botched attempt to rescue the American hostages held in Iran, which lost him re-election in 1980. Conversely – in a very few cases – a crisis may come at the very moment that the politician might have chosen if a genie had granted him or her a wish. And so it proved for Margaret Thatcher in 1982. With her government lagging behind in the polls and facing serious internal problems, the Falklands War came at just the right time for her. Any earlier and the 'iron' in her soul might have been overlooked and she might only have been thought 'very determined'. Instead, in the flames of military crisis a reputation was forged in 'blood and iron'. Armoured with the justice of her cause she became the stuff of legend. The Falklands War created the Thatcher legend – yet it was a war that should never have been necessary in the first place. And when fought it should have been lost, despite the extraordinary courage and efficiency of Britain's armed forces.

The Argentinian invasion of the Falkland Islands in April 1982 was a national disgrace. A British government had not been surprised by a 'sneak' attack or even 'stabbed in the back' by an erstwhile friend. Instead Argentina had got tired of threatening to take the 'Islas Malvinas' and had simply walked in through the front door, when the British had left the house unattended. Britain had made light of Argentine 'sabre-rattling' and had ignored an invasion threat as recently as 1977. Thus when war came in 1982, it was the result of political errors of awesome dimensions.

It must never be forgotten that the Margaret Thatcher who was to gain so much from the successful campaign to retake the Falklands was the same Margaret Thatcher whose government had totally misread Argentinian intentions and had let the islands be taken in the first place. Although the British people may have regarded the crisis of April 1982 as a storm that had suddenly blown up, the Falklands had long been a subject of vital national importance to Argentina. Except for experts at the Foreign Office, few people in Britain knew how strongly Argentina had asserted her claim to the islands, particularly since 1945. There had been numerous crises in the 20th century, during which Argentina had claimed sovereignty: notably in 1927–8, 1933, 1966 and 1976. In 1977 there had been a serious threat of an Argentinian invasion of the Falklands. A British ship had been fired on and a Royal Navy submarine had been sent to the South Atlantic as a deterrent by the then Labour government. Yet by February 1981 it seemed that negotiations would settle everything. But the overthrow of General Viola and his replacement by an unpopular military junta headed by General Galtieri changed all this. Renewed 'sabre-rattling' soon caused a breakdown in negotiations. Faced with this new threat the British Foreign Secretary, Lord Carrington, and Prime Minister Margaret Thatcher, had to decide whether the situation required the presence of another deterrent force in the South Atlantic. In the event, they chose not to send one for budgetary reasons and because they did not take the Argentinian threat seriously. Not for the first time the Treasury was dictating Britain's foreign policy.

Intelligence sources did not agree with Thatcher and her ministers. They made it clear that the danger of an Argentine invasion of the Falklands was far greater than in 1977 and that some response by Britain was imperative. Yet even as late as 29 March 1982, Lord Carrington was still adamant that Her Majesty's Government was reserving its military options and hoping for a diplomatic solution. Secretly the nuclear submarine HMS *Spartan* had been ordered to the Falklands, but it was already too late. Britain was about to be fatally compromised by a government that had got its priorities wrong and was about to pay a high price in blood and treasure.

Margaret Thatcher's policy of deterrence in the South Atlantic failed to prevent the Argentinians from invading the Falklands, and the root of this failure was to be found in a series of incorrect strategic assumptions. Perhaps the most important of these was the belief that the Argentinians would not resort to force to take the islands. The dispute was one of the longest running in the history of international relations, dating back to the 1770s, and Argentina had never previously resorted to arms. There had been threats but these were regarded as empty ones, products of the rhetoric of South American politics. 'Sabre rattling' about the Malvinas seemed to be entirely for domestic consumption in Argentina. Yet the Junta that ruled in Argentina was a new and unpredictable factor in the equation. The Foreign Office should have left Britain in a position to react to force if the traditional policy of deterrence failed.

Britain made a second mistake in underestimating the significance of the Malvinas for Argentina. The islands may have been a very distant colonial acquisition for Britain, but to the average Argentinian they represented a part of the homeland, whose importance was instilled in every child at its mother's breast. Such intensity of feeling is generally an indication that a problem will not just 'go away'. British bluff and stalling were hardly going to deter an enemy who felt as strongly as Argentina

British prime minister Margaret Thatcher points to the sign of welcome displayed by 'Y' Company, 1st Battalion, the Royal Hampshire Regiment, when she visited them at Goose Green in January 1983.

did. Britain's feeble and half-hearted diplomacy never suggested to Galtieri that Margaret Thatcher's government cared 'tuppence' for the Falklands. And if Britain really did care – as Thatcher claimed – why did she not tell the Argentines to keep off and support her policy with some military muscle?

Ideally, Britain might have liked to commit adequate forces to the defence of the Falklands, but the economic realities of the 1980s made this unthinkable to Margaret Thatcher's government. Like previous Conservative prime ministers she preferred to wear the trappings of empire without updating the costume. By failing to equip an adequate force in the South Atlantic she misled the military Junta in Buenos Aires into believing that Britain did not care about the Falklands enough to defend them. Galtieri assumed that an invasion would provoke no more than a token British reaction. When Britain reacted as if the Isle of Wight had been occupied by French fishermen, it seemed so extreme to the Argentinians as not to be worth taking seriously. It was this lack of mutual understanding that turned a 'war of words' into a bloody struggle that cost a thousand lives.

The Junta's failure to realize that Britain would fight over something about which she seemed not to care was quite understandable. Britain had already decided to withdraw the ice patrol ship HMS *Endurance* from Falklands waters and only the occupation of South Georgia by Argentinian scrap metal dealers persuaded the Admiralty to let her remain. Otherwise, Britain never made Argentina aware of the possible risks she ran if she resorted to military means. Britain, in fact, had substituted a policy of bluff for a real deterrence. When her bluff was called, she was forced to respond on a scale and at a cost much higher than would have been needed to maintain effective deterrence. The presence of a nuclear submarine in Falklands waters, and a clear message to the Argentinian Junta that it was there and would be used if necessary, would have saved both countries from a 'shooting' war.

In fact, the Argentinians had chosen quite the wrong time to escalate the crisis. Britain was in the process of running down her naval capability. The aircraft carriers *Hermes* and *Invincible* were due to be retired and sold to Australia respectively, thereby removing Britain's capacity to operate sea-based airpower. Moreover, plans for future carriers had been scrapped. And the retirement of the last Vulcans would remove Britain's long-range strike capacity completely. Another year or 18 months would have left Britain incapable of launching an amphibious operation to land troops on a distant island. So Argentina was justified in thinking that the whole tenor of British military thinking was to reduce commitments and concentrate on a European scenario in which her role was to provide an anti-Soviet, anti-submarine force. Britain herself probably felt her 'gun-boats' were safely stowed away with 'Drake's Drum' in historical mothballs – and she was making no secret of this.

And then, from the ruins of Britain's strategy, arose a champion, playing the rôles of Lord Chatham, William Pitt the Younger and Winston Churchill – without even having auditioned for the parts. No South American junta was going to kick sand in Britain's face and get away with it. She might have botched the strategy but just watch the tactics. Following in the footsteps of her mentor – Sir Winston – she was going to do her best to commit her country to an amphibious operation in the best traditions of Gallipoli, Narvik and Dieppe. Who dared to tell her it could not be done – that Britain would be outnumbered ten to one in the air, that she had no airborne early warning system and that frigates and destroyers would have to be sacrificed to make up for the deficiency? Who even hinted that without substantial American aid the Royal Navy could no longer police the globe as it once had? She had no ears for fainthearts or experts. She would *will* it to happen. And it did happen – but it owed little to her. British professionalism, American aid and astounding luck all combined to bring about the unlikeliest victory since the battle of Midway. Never mind the cost – it was the result that counted. She had demanded the impossible and it had been done. She was the Iron Lady, forged in the flames of war.

U-turn at Dunkirk

Germany began to lose the Second World War on 24 May 1940. At the height of her triumph over France there occurred an incident that has always baffled students of military history. At a time when the British forces in France were being pressed back against the coast and were facing certain defeat, the Germans called off their attacks long enough for the British troops, and a substantial number of French ones, too, to be evacuated from Dunkirk. Someone had taken a decision unprecedented

Tanks and equipment abandoned on the road to Dunkirk during the Allied retreat of May 1940. Hitler's decision not to wipe out the British forces in the town and on the beaches has long perplexed students of military history.

in military history – to allow a beaten enemy to escape to fight again. The assumption has always been that the decision was Hitler's, and that the military dilettante was up to another of his tricks. But is this entirely fair?

The truth was that Hitler, after his extraordinary successes against France, was beginning to lose his nerve. Could things really continue to go so well? Having gained such a prize, was it about to be taken away again? On 17 May, General Franz Halder reported, 'Führer is terribly nervous. Frightened by his own success, he is afraid to take any chance and so would rather pull the reins on us.' A few days later Halder continued, 'Führer keeps worrying about the southern flank. He rages and screams that we are on the way to ruin the whole campaign. He won't have any part in continuing the operation in a westward direction.' Hitler's nerve had been broken not by failure but by success.

On 24 May the famous 'Halt' order reached Guderian's advanced forces as they were crossing the Aa Canal, just prior to closing in on Dunkirk. Guderian's feelings can best be left to the imagination. As the supreme exponent of *blitzkrieg*, he knew that you did not stop once you had the enemy on the run. That was the whole point of it. But as the 'Halt' order came direct from the Führer, it must be obeyed. And so the German tank crews sat and whistled as the British moved towards the Dunkirk beaches and the evacuation boats assembled. It was a curious kind of war. Yet overhead the Stukas were still bombing the retreating allies, so why had the tanks been stopped?

Several theories have been advanced to account for Hitler's decision but most can be simply dismissed. One idea was that the Germans were afraid to see their tanks bogged down in the Flanders mud, as in the First World War. However, there is no evidence that any German generals feared this, and it is unlikely that this is what motivated Hitler. A second view is that Hitler was offering the British a chance to escape so that they would be more willing to make peace with Germany. Although Hitler did admire the British, on occasions, and sometimes wished for peace with them, this theory does not really hold up as he was quite prepared to allow the Luftwaffe to continue to bomb British forces at Dunkirk. Certainly Goering did try to convince Hitler to allow the Nazi-orientated Luftwaffe to steal the thunder from the more politically 'traditional' generals. But the most likely reason for the 'Halt' order has already been hinted at in Halder's diary entry: fear of the southern front. Hitler – feeding on the fears of Rundstedt – was genuinely alarmed at the possibility of German forces being struck by the substantial French forces to the south. He did not want to see his tank strength so diminished by the fighting around Dunkirk that it was too weak to resist a new French challenge. In simple terms, he got cold feet.

But the effect of seeing the entire army forced to capitulate along with the French might have been all that was needed to crack British morale. Even Winston Churchill might have found it difficult to make capital out of such a dismal defeat – it would have been the biggest in British history – rather than the epic evacuation from Dunkirk, which could be embroidered in such a way as to seem more than a victory.

If Hitler was influenced by the fears of some of his more conservative generals, notably Rundstedt, that cannot conceal the fact that the final decision was his, and his alone. It is not overstating the case to suggest that it was his – and Germany's – biggest blunder of the entire war. Hitler had proved that rather than taking his place among the great conquerors of history – the likes of Alexander, Caesar and Genghis Khan – who would hardly have missed the chance of destroying their most dangerous opponent if they had him in their clutches, he was nothing more than a 'Sawdust Caesar'.

A woman's place

During the First World War women had made such a contribution to the war effort in each of the combatant nations that it might have been taken for granted that, when war came to Europe in 1939, their services would be called on again. In Britain women workers were welcomed with open arms and later conscripted. In Nazi Germany, however, things were very different. Hitler was adamant that a woman's place was in the home and not the workplace. From the early days of Nazism women had been barred from executive posts, and by 1936 even female lawyers and schoolteachers had lost their jobs. Hitler equated the well-being of the Reich with the size of the birth rate, and he encouraged women to have babies as their contribution to the military strength of the state.

When war broke out in 1939, women were vastly under-employed in Germany.

Hitler's highly traditional view of the role of women can be clearly seen in this propaganda poster for the Nazi relief organization *Mutter und Kind* ('Mother and Child'). Germany's failure to employ women in war production was a grave error in the era of 'total war'.

Although many still worked in farming or domestic work, at least five million were available for war work but were forbidden to do it. As casualties on the Eastern Front began to mount Germany had no alternative but to use slave labour in the factories because of the shortages of civilian manpower. With women working in British industries producing war materials, Germany found herself falling behind badly in the production of tanks, planes and munitions. The need to maintain huge armed forces in 1942 and afterwards meant that there was always a shortage of labour. Albert Speer told Hitler that if he were able to use women in the factories he would be able to release at least three million more men for the army. But the Führer was adamant. Military needs would not be allowed to distract him from Nazi ideology. And so the German armies suffered while the women stayed at home. A vital military mistake had been made, thanks to the narrow intellect of an ideological fanatic.

The mother – and father – of all defeats

There is little evidence to suggest that gangsters make good generals, that street fighters make good soldiers or that bluff is a substitute for strategical thinking. Any such evidence might lead one to apply the title 'Sawdust Caesar' more tentatively to Saddam Hussein. Yet the results of the Gulf War of 1991 suggest that Saddam applied the methods of the terrorist, the mass-murderer, the hijacker and the gambler to a military situation and ended up with the 'Mother of all defeats'. In military terms Saddam did everything wrong. Whatever might have been the qualities of his generals we will never know, because Saddam tied their hands by his bizarre strategy. Like some egomaniac fly, he defied the world to swat him – and it did just that.

During the 1980s successive American governments overlooked Saddam Hussein's human-rights abuses within Iraq because they saw him as a useful bulwark against the perceived threat from fundamentalist Shi'ite Iran. They supplied him with military hardware, and trained Iraqi personnel in the use of modern American equipment. Presidents Reagan and Bush believed that Saddam's extreme behaviour could be moderated, and that he would always be susceptible to bribery in the form of military hardware because he was essentially a 'tin-pot' dictator – a man who enjoyed holding military parades and pretending he was an Islamic hero, a new Saladin. But they were underestimating him. Saddam was an opportunist and a man who regarded moderation as a weakness to be exploited. He built up his army not just to play with, but to use against weaker neighbours. He convinced himself that he had nothing to fear from the United States who, he believed, for all its military might, had not got the guts to fight a real war. His study of history was selective. Vietnam, as well as more recent examples in Iran, Lebanon and Grenada, showed that the Americans did not have the stomach for casualties, something which he knew the Iraqi people were willing to suffer if he asked them to. Thus, no amount of diplomatic pressure could worry Saddam. In the final analysis he believed no American president would think a place like Kuwait worth the bones of a single US Marine.

Even after the American-led coalition had been ranged against him, Saddam did not think the Americans really meant to fight. He confidently boasted to US Secretary of State James Baker, 'Your Arab allies will desert you. They will not kill other Arabs. Your alliance will crumble and you will be left lost in the desert. You don't know the desert because you have never ridden on a horse or a camel.' Saddam's confidence seemed complete. But had he got the military muscle to back him up? The answer was, possibly, 'yes'.

Like other 'Sawdust Caesars', Saddam Hussein did not like his ministers to disagree with him. The intelligence they fed him was simply designed to support his own beliefs. They knew that nobody who told Saddam too much of the truth would survive long. Saddam himself had given some thought to Western psychology. As well as believing the USA would not accept high casualties, he also believed that he could use the threat of chemical and biological weapons to be delivered by his Scud missiles against Israel as a form of blackmail. Westerners would be horrified at the thought of such weapons being used and would do anything to avoid it happening. With no democratic electorate to whom he was answerable, Saddam Hussein clearly felt he could be as 'bad' as he liked and nobody would dare do anything about it.

In addition, since the war with Iran, he had spent so much money on state-of-the-art defences against air attack, including bunkers so strong and sophisticated that they could resist even the latest American technology, that he felt confident in outfacing Western threats from the UN. Nor was Saddam slow to exploit Western intelligence failures. He encouraged the West to overestimate his military potential – one US report stated, 'Iraq is superb on the defense.' Thus by the UN deadline of 15 January, by which time he was supposed to have evacuated Kuwait, the Americans believed he had 540,000 men dug into a formidable line of defences on the Kuwait–Saudi border. In fact he had scarcely more than 260,000. Saddam thought he could bluff the Western powers because they lacked the spirit to put him to the test. But for once his brinkmanship failed and he had to demonstrate military leadership rather than bluff. Was he up to it?

When the Americans and their allies called Saddam's bluff by launching Operation Desert Storm, the Iraqi dictator was revealed for what he was, a posturing bully. He simply abandoned his front-line units in defensive positions, giving them no air cover whatsoever. The best of the Iraqi air force took sanctuary in Iran, while the first two Mig-29s that took off to engage American F-15Es only succeeded in shooting each other down by mistake. Meanwhile day and night, for over five weeks, the devastation of Iraq by precision bombing continued.

Saddam's strategy baffled military experts. He did not use his air force at all and abandoned his ground troops to endure an unbelievable pounding. The Iraqi front-line soldiers received no food or water. One platoon was only saved from dying of thirst by the heavy rain that fell. Saddam's random use of Scud missiles against Israel was a political rather than a military tactic. Its aim was to provoke the Israelis into retaliating against Iraq, in the hope that the spectacle of an Israeli attack on Iraq would cause states such as Syria and Egypt to leave the international coalition ranged against Iraq. He did not use the terror weapons that he had used on the Iraqi Kurds because

A giant propaganda poster in Baghdad depicting Saddam Hussein as the champion of the Arab world. Much of Saddam's policy towards the West was based on bluff, but when his own bluff was called in the Gulf War of 1991 he suffered a shattering military defeat.

he feared that he would be toppled from power and put on trial by the Allies if he did. In fact, he displayed a passivity that was almost disconcerting. Few generals in history have ever been prepared to take such a beating without either striking back or surrendering to reduce casualties. But Saddam, who seemed to care nothing for the welfare of his people, allowed the war to go on, thereby defying the Western powers with a display of bravado that seems close to the hearts of some Arab politicians. One is reminded of President Nasser's defiance of Britain and France in 1956 and the political advantage he gained from military defeat.

Saddam Hussein had been right about one thing: the Americans were afraid of a heavy 'body count' in the struggle to liberate Kuwait. They had 17,000 body-bags in Saudi Arabia, but were hoping to get away with fewer casualties. In the event, their casualties were fewer than on many NATO manoeuvres. But what Saddam took for fear was a natural aversion on the part of a democratic state to suffering casualties. To a bully like himself fear only meant weakness. But by 25 February even Saddam Hussein had had enough and ordered a general withdrawal from Kuwait. Militarily, the war had been a 'no-contest walkover', in the words of one expert.

Saddam Hussein most resembles Mussolini as a 'Sawdust Caesar'. Unlike Hitler, whose bluff was generally backed by military strength, both Mussolini and Saddam fell victim to their own rhetoric. Each was able to deceive gullible democratic politicians but, as an American president once said, 'you cannot fool all the people all the time'. Saddam's gestures – horrible as they might be – came from the same stable as Mussolini's. Saddam's were darker in hue, less imbued with the operatic tendencies of the Italian, but they were equally hollow. Just as Mussolini brought down on Italy catastrophic defeats in North Africa, Somalia and Greece, so Saddam challenged the United States and suffered the 'mother and father' of all military defeats.

Part II

THE BATTLE OF CANNAE (216 BC)

The battle of Cannae was undoubtedly Hannibal's masterpiece. Indeed it seems almost churlish to speak of the incompetence of his opponents in the face of a master's supreme skill. Yet Hannibal was blessed by facing a Roman army enormous in numbers yet divided in command. It was this very system – which allowed the two consuls to alternate control on a daily basis – that condemned Rome to the heaviest defeat in her history.

During his campaign in southern Italy in 216 BC, Hannibal depended for supplies on living off the country or on capturing important food depots. It was this factor that rendered him most vulnerable to defeat. In the spring of 216 BC he had marched south, crossing the River Aufidus, and occupying the town of Cannae, which was an important grain depot for the Romans. Cannae was situated on a hill overlooking a very broad plain across which the sluggish River Aufidus flowed to the sea. Hannibal knew that by seizing Cannae he was not only securing food for his troops, but threatening Rome's supplies to her own army.

At this stage the Roman army was still commanded by the previous year's consuls, Servilius and Atilius. They were not eager to seek battle with Hannibal so near to the end of their consulship. They preferred to wait for the new consuls, Paulus and Varro, and leave the decision of whether to fight or not to them. Nevertheless, the Roman Senate was looking for an opportunity to fight a decisive battle against the Carthaginians to avenge the defeats of previous years. According to the historian Polybius the Romans fielded eight legions in 216 BC, numbering some 40,000 men, supplemented by an equal number of auxiliaries. If this is true, it makes it the largest Roman army ever raised up to that time. In fact, it was far too large to be commanded by generals of that era; the confusion that ensued in the battle was a result of overloading the simple command structure then in existence. Hannibal, on the other hand, had an experienced staff system and very able subordinate commanders. He knew he could trust men like his brother Mago, and officers like Hasdrubal, Hanno and Maharbal to control his 40,000 foot and 10,000 cavalry. It gave him an inordinate advantage when the time for action came. In spite of the enormous Roman advantage in infantry – far more than could be brought to bear at any one point – Hannibal's cavalry was more numerous and more powerful than that of the Romans. In another area – that of experience – Hannibal had an even greater advantage. His men were generally veteran soldiers who had been with him on numerous campaigns. The Romans, on the other hand, now under the rotating command of Paulus and Varro, were mainly new levies without battle experience.

As Hannibal's strength rested with his cavalry, he was anxious that any battle should take place on the wide plain, on the west bank of the Aufidus. The Romans, naturally, would have preferred to fight under the hill of Cannae on the east bank. The day the Roman army reached Cannae was one on which Paulus was in command and he suggested to Varro the advantage of crossing the river and assembling on the east bank, which would give Hannibal less chance to use his powerful cavalry. Varro disagreed, and the following day recrossed the river to face Hannibal on the west bank. He knew what was expected of him and the army – an immediate and decisive victory – and was not the sort of cautious and thoughtful commander who might have manoeuvred to gain an advantage. Instead he relied on weight of numbers alone. So stubborn was Varro that every time Paulus suggested fighting on the hilly ground to the east it only made him more determined to do the opposite. After further crossings and recrossings of the river the Roman army drew up in formation on the west bank, some two miles away from Hannibal's army. In the June sunshine the new legions – unaccustomed to the burning heat of southern Italy, and weighed down by the weight of their armour and their weapons – were suffering far more than Hannibal's Spaniards and Africans.

When the command reverted to Varro, the latter, abandoning his previous preference for the west bank, ordered his troops to leave camp just after dawn and cross over to the east bank of the river in order to threaten Hannibal's food supplies from Cannae. If Varro had hoped to take Hannibal by surprise he was disappointed. The Carthaginian leader had been waiting for just this moment: the chance to destroy the Romans once and for all. He swiftly crossed the river and drew up his highly disciplined troops in an unusual formation – one that was eventually to become the most famous in all military history. On his right flank were the Numidian horsemen whom he knew he could rely upon to scatter the Roman cavalry. On his left, alongside the river, he placed the heavy cavalry of the Spaniards and Gauls. But it was in the middle, where he himself took command, that the battle was to be won. Most of his Spanish and Gaulish infantry was massed in the centre, facing the Roman legionaries whose lines stretched back for hundreds of yards. Through skilful manoeuvring Hannibal had secured the advantage of both sun and wind; the Romans suffering the unpleasant effect of both the burning sunlight and the sand whipped into their faces by the scirocco.

As the Carthaginians began to advance it was apparent that their infantry centre was drawn up in a convex shape. But the early exchanges were between the rival horsemen. The Spanish heavy cavalry routed their Roman opponents and quickly drove them off the plain. On the other wing the Numidians had already routed the Roman auxiliary cavalry. With the latter fled Varro, the consul who had foolishly sought this battle.

In the centre Hannibal's infantry was gradually giving way, as the Romans began to flatten the Carthaginian line. Almost imperceptibly the convex shape had become concave or U-shaped. Sensing a breakthrough, the Roman commanders urged on the massive wedge of legionaries. But as the Spaniards and Gauls gave way in the centre, the two wings of African troops began to curl round the Roman flanks. It now became apparent that the massed legionaries in the centre were being enveloped by the Africans. At the same time the Carthaginian cavalry, returning from their victory over the Roman horse, now struck the Roman legionaries from the rear.

Hannibal could see that the moment had arrived to spring the trap, and he ordered a trumpeter to make the signal. At once the Africans completed their double envelopment, by which the huge mass of Roman legionaries was completely surrounded by the Carthaginian army. It was a tactical masterpiece, and one that has inspired generals right through the ages. The battle now turned into one of the most terrible massacres in history. The killing at Cannae that day exceeded even that of 1 July 1916, when the British were slaughtered on the Somme. Never before had the Romans raised so many men for battle, but now they were trapped so tightly that only those at the extreme edges could even lift their weapons or reach their opponents. Hour after hour passed and the killing went on. So vast was the death toll that ancient writers could hardly contain their disbelief. The lowest figure – too low in view of the impossibility of escape for the Roman footsoldiers – was a mere 20,000 killed, while Appian and Plutarch speak of 50,000 dead, Quintilian 60,000 and Polybius 70,000. Modern historians have learned to doubt the fantastic figures for armies and casualties given by medieval chroniclers, but it is implausible that casualty figures for this battle could be much lower than 40,000 killed, and they were probably far higher.

Of Hannibal's genius there can be no doubt, but one is forced to the conclusion that the split command of Paulus and Varro contributed significantly to the Roman defeat in view of their obvious incompatibility as commanders. Facing such a skilled opponent, Paulus felt inclined to adopt a more cautious approach, while Varro, presumably underestimating Hannibal or having too much confidence in mere weight of infantry numbers, wanted to press for immediate action. Hannibal might well have defeated Paulus as totally as he had defeated Varro, yet one suspects that few Roman generals would have fallen so completely for his tactics. Varro was a headstrong fool. Leading with his chin he was always vulnerable to a counterpunch. The bunching together of the legionaries was poor generalship – once the first two or three ranks were engaged there was absolutely nothing for the rear ranks to do other than wait until the men in front of them were killed. Piling more and more men into the centre was a dangerous tactic once the Roman cavalry had been defeated. But for all Varro's blunders nothing should detract from Hannibal and his strategic masterpiece on the plains of Cannae.

THE BATTLE OF CARRHAE (53 BC)

Ambition without ability can be a dubious blessing. As Roman senator Marcus Crassus sat through triumph after triumph by his fellow Triumvirs, Caesar and Pompey, he wished for nothing better

Marcus Licinius Crassus is cut down by the Parthians as he attempts to parley after the battle of Carrhae, 53 BC. The Parthian victory at Carrhae stopped the Roman invasion of Parthian Mesopotamia in its tracks, and was a devastating blow to Roman prestige in the east.

than a victory to match any of theirs. Unfortunately he chose as his battlefield the notoriously difficult land of the Parthians, where many Roman armies – led by more able generals than he – had come to grief in the past. But ambition drove him onwards and blinded him to the advice of seasoned campaigners. His lack of military experience caused him to overlook the difficulties of the terrain he had to face and underestimate the dangers of facing the powerful Parthian horse archers. These mistakes were to cost him his life and Rome one of her worst defeats in the east.

Around a well-stocked dining table, with convivial friends as guests, the problems of the campaign appeared to Crassus to have been overrated. He boasted to his friends that his only real problem would be 'the tediousness of the march and the trouble of chasing men who dare not come to blows'. This probably made more sense to men in their cups than to sober men in the cold light of day. But Crassus clearly felt that he knew best; when the Armenian king told him that the best way of campaigning against Parthia was to move through the hills and high ground to reduce the effect of the Parthians' cavalry, Crassus laughed. What could an Armenian tell a Roman about war? Rejecting the king's advice he chose instead a desert route for his army of 36,000, made up of seven legions and 4,000 cavalry. But marching straight through enemy country made foraging difficult and rendered his own supply lines vulnerable. When Cassius, one of his generals, advised him to rest in one of Rome's frontier cities to give his troops time to acclimatize – a course of action that would also allow him to reconnoitre the Parthian positions and plan a route of advance – Crassus was furious at having his decisions questioned. Disdaining Cassius's idea of following the Euphrates towards the Parthian capital of Seleucia, to ensure a constant supply of water and to protect one of his flanks, Crassus deliberately ordered an advance away from the river and straight out into the desert. He seemed inspired by a childish urge to do the opposite of what anyone suggested. As the legionaries trudged through the deep sand they began to be hit by 'Parthian tactics', by which horsemen would dart in and out, firing over their shoulders with impressive accuracy, and always getting clear before the Romans could respond. These 'flea bites' did not worry Crassus; such tactics only confirmed his belief that the Parthians were cowards who were afraid to face the Romans in a fair fight. Yet the Parthians' hit-and-run attacks were taking their toll on the legionaries, who were suffering casualties without being able to respond.

As the pressure from the Parthian commander, Surenas, increased, Crassus formed his men into a defensive square and continued his advance in this formation. But Roman progress through the desolate wasteland was now pitifully slow, and although they outnumbered the Parthians by more than three to one, they could never drive them away for long. Crassus's ignorance and underestimation of the enemy revealed a prejudice that he was soon to regret. The appearance of Surenas – dressed effeminately (in Roman eyes), wearing make-up and with his hair parted like a woman – prompted Crassus to ridicule him openly and call him a catamite. This served to anger the Parthians and made it certain that when the time came they would show Crassus and his men no mercy.

In an attempt to drive the Parthians away from the slowly advancing Roman square, Crassus sent out his son, Publius, with a force of cavalry and archers. At first, Publius achieved some success with his Gaulish light cavalry, who outmanoeuvred the heavily armoured Parthian cataphracts. But the Parthians drew Publius away from the main Roman force and surrounded his troops, wiping them out and killing him. They then cut off Publius's head and displayed the gory object on a lance for his father to see.

This personal tragedy had a sobering effect on Crassus. He realized that he had no choice but to withdraw, abandoning 4,000 wounded men to be slaughtered by the Parthians. But during the confused retreat, much of it conducted by night, the Roman army began to break up, with individual officers leading their troops back to their fortified camp. Cassius managed to lead as many as 10,000 men to safety, but the bulk of the Roman army was killed when the Parthians overran the camp at Carrhae and treacherously murdered Crassus during a parley.

Crassus had no redeeming features as a general. He had behaved like a fool throughout the campaign; indeed few disasters in military history were so much the responsibility of the commanding officer. He made every mistake possible – underestimating and ridiculing the enemy, making light of the difficult terrain and the obvious need for water, ignoring the sensible advice of men who had campaigned in Parthia and allowing his enemy to dictate tactics to him. Roman losses at Carrhae were severe, with 20,000 men killed and 10,000 captured and enslaved. It was one of Rome's heaviest and most ignominious defeats.

THE BATTLE OF MYRIOKEPHALON (1176)

The difference in military tactics between the Byzantines and the Seljuk Turks resulted in a number of significant military disasters – mainly for the Byzantines. Two centuries of experience of Seljuk warfare should have enabled them to avoid situations in which the Turks were able to exploit their mobility to the disadvantage of the more ponderous Byzantine military machine. Thus disasters when they did occur were almost exclusively the fault of the Byzantine commanders. The defeat suffered by the Emperor Manuel Comnenus at Myriokephalon was the result of both arrogance and stupidity on his part.

In 1176 Manuel had decided to solve the Turkish problem once and for all, by liberating Anatolia (Asia Minor) from the Seljuks. First he sent an army

A stone relief depicting Seljuk warriors of the 13th century. The Seljuk defeat of the Byzantine Emperor Manuel Comnenus at Myriokephalon in Phrygia in September 1176 ended Byzantine hopes of expelling the Seljuks from Anatolia.

under his cousin Andronicus to regain control of Paphlagonia (in what is now northern Turkey, on the Black Sea), while he raised an enormous imperial force, swelled by mercenary troops from Europe and Asia, and weighed down by a vast array of siege engines. He set off from Constantinople intent on marching on the Seljuk capital at Konya. The Seljuk leader, Kilij Arslan, hearing of Manuel's enormous power, sent to ask for peace but Manuel, convinced that the Turk could not be trusted, refused to negotiate.

In September, however, the expedition to Paphlagonia was disastrously defeated and the head of Andronicus was sent to the sultan. News of this setback and of the death of his kinsman only made Manuel more determined and he pressed on into the mountains of Cappadocia. His progress was constantly interrupted by accidents to the huge machines and towers that were being laboriously dragged along with the army. It was a quite inappropriate force to be seeking battle against the mobile Turkish horse archers, as generations of Byzantine generals could have told him.

Manuel reached the pass of Tzibritze, at the far end of which stood the ruined fortress of Myriokephalon. It was an obvious place for an ambush and – true to form – Kilij Arslan had planned to attack the imperial army as it passed through. So certain were the Byzantine generals that the Seljuks would attack them there that they begged the emperor to be more cautious. But he refused to listen to them and paid more attention to the foolish young princes and knights of the Byzantine household cavalry, who were spoiling for a fight. As a result he allowed his unwieldy force to enter the pass even though the Seljuk army was clearly visible on the hillsides.

The Byzantine vanguard succeeded in forcing its way through the pass, holding off Seljuk attacks on its flanks. But as the main body of Manuel's army followed down the narrow path, Turks attacked it from all sides. The result was total confusion, with the huge towers and machines toppling over and blocking the way forward as well as the way back. Baldwin of Antioch, brother-in-law to the emperor, led his knights into the hills against the Turks but was overwhelmed and killed. As the Byzantine soldiers, trapped on the narrow road, saw the defeat of their knights, many of them panicked and began to flee. But so tightly packed were the soldiers on the road that few could even lift their arms to wield their weapons. Manuel did nothing to try to rally his troops and fled down the valley, being one of the first to escape. Seeing their leader flee, the whole Byzantine army now tried to follow him, causing unbelievable chaos and destruction. The Turks scarcely had to strike a blow as the Byzantines seemed intent on destroying themselves.

When darkness fell the sultan sent a messenger to Manuel, offering him peace on condition that he withdrew his army and dismantled the new fortresses he had been building in Anatolia. Manuel could scarcely believe his luck. Kilij Arslan must have no idea of how great a disaster the Byzantines had suffered. Admittedly their vanguard was still intact, but the main body was all but destroyed. Kilij's decision to let Manuel go was an unpopular one with his people – even perhaps a militarily unwise one – yet Manuel knew in his heart of hearts that he had been released from the jaws of death and that the military power of the empire was broken. As at the great defeat at Manzikert a hundred years before, the Byzantines had squandered every material advantage and – through sheer stupidity in underestimating an enemy they knew well – had suffered a disastrous defeat.

THE BATTLE OF STIRLING BRIDGE (1297)

War is an expensive business, but how many commanders would welcome a treasury official at their shoulder throughout a campaign, reminding them of just how much everything was costing. Yet such was the fate of John de Warrenne, Earl of Surrey, the Warden of the North, as he marched towards Stirling in 1297 at the head of an English army. His task was to suppress the rising of the Scots under William Wallace, and with him rode the English Treasurer, the unpopular Hugh de Cressingham. The task was not much to Surrey's liking. An elderly gout-ridden man in his late sixties, he would have much preferred to spend his days on his rich English estates rather than chasing the wild and barbaric Scots, not one of whom was worth a ransom. Cressingham, a Falstaffian figure with none of the humour, was universally hated by the Scots. He drove Surrey on, much against his better judgment, constantly telling him that the sooner the campaign was over, the sooner the English army could be paid off. It became Treasury policy to fight Wallace where and when the English army could find him, and to cut costs Cressingham had already sent the soldiers of Henry Percy's substantial division back to their Northumbrian homes.

Wallace, showing a far greater strategic sense than Surrey and Cressingham, had taken up a strong position near Cambuskenneth Abbey, threatening Stirling Castle. Between him and the English was the fast-flowing River Forth. Now rivers are merely one of many kinds of natural obstacles that any commander has to face. To cross a broad stream he will normally need to find a ford, but to attempt a crossing in the face of an enemy on the opposite bank is an action needing the most precise planning. The existence of a bridge does not necessarily make things easier, as Warrenne was to discover as he approached Stirling on 11 September.

William Wallace had hidden his rag-tag army in the wooded hills overlooking the river as it flowed past Stirling. A single wooden bridge, wide enough for just two men abreast, crossed the river near the abbey and led to steep slopes on which Wallace

The battle of Stirling Bridge, 11 September 1297, in which the Scottish rebel leader William Wallace trapped an English army attempting to cross the River Forth by a narrow wooden bridge. Wallace went on to devastate England's border counties.

himself was stationed. As Warrenne approached with his troops, he had already determined to cross by the bridge and attack the Scots' position, even though he had not carried out any reconnaissance and knew little of the difficulties his men would encounter. In fact he never for a moment wondered why the bridge had been left standing by the rebels. Could it be that they wanted him to cross?

The idea of moving thousands of men across a narrow bridge in the face of the enemy was an extremely ill-conceived one. It was pointed out to Surrey that it would take eleven hours to move all the men across, during which time anything could happen, but Cressingham continued to urge him on saying, 'There is no use, Sir, in drawing out this business any longer and wasting the king's revenues for nothing. Let us advance and carry out our duty as we are bound to do.' When Sir Ralph Lundy told Warrenne that he knew of a ford less than a mile away, where the men could cross 30 abreast, the old warrior refused to listen. His low opinion of the Scots as warriors was well known and 'the wiser heads in the camp were filled with dismay at a resolve inspired by a foolish and overweening contempt for the enemy'.

The English vanguard, led by Sir Marmaduke Twenge and Cressingham himself, now began to cross the bridge and form up on the northern bank of the river, overlooked by Wallace and his soldiers. Showing strong discipline and patience, the Scots waited until at least a third of the English had crossed before attacking. Wallace's men, some armed with spears, others with the lochabar axe, now charged down the slopes, seizing the bridgehead and stopping the main part of the English army from crossing, while Wallace attacked the trapped vanguard. Although Twenge was able to fight his way back, Cressingham, a hundred knights and several thousand English footsoldiers were massacred on the river bank.

It had been an act of madness on Surrey's part to walk into such an obvious trap, but one can perhaps pardon a tired old man, urged on by a civil servant who knew little of military tactics and whose contempt for the enemy led him to underestimate them. Cressingham paid for his folly with his life. After the battle his bloated body was stripped and flayed, with pieces of his skin being carried away from the battlefield as tokens. Surrey made little effort to rally his men and rode hard until he had reached the security of Berwick.

THE BATTLES OF GRANDSON, MORAT AND NANCY (1476–7)

Charles the Bold of Burgundy – the 'Rash' is a far more appropriate translation of the original French – was one of the grandest rulers of the 15th century. Enormously rich and excessively ambitious, Charles lacked the abilities to achieve his goals, which were little less than to make himself a king and his duchy a kingdom. To achieve this he knew that he would need a powerful army. Charles' desire to gain military renown acted as a spur to him, particularly during the 1470s. Charles saw himself as a new Alexander, a greater Caesar, a more glorious Hannibal. His pride often made him seem ridiculous and his aspirations to sit alongside kings and emperors involved his duchy in excessive taxation to feed his insatiable appetite for war. The problem was that he was one of the worst generals of his day who never recognized when he was outmatched, notably in his last three dreadful encounters with the Swiss. A more appropriate name for him might have been Charles the Mad, for his military judgment often bordered on the insane.

For his campaigns against the Swiss, Charles decided to rely mainly on a mercenary army. He worked on the principle that if he selected the best of each nation – and of each kind of military expertise – he must end up with the best possible army. He therefore hired English archers, German handgunners, Italian light cavalry, Flemish pikemen and Burgundian knights. Far from producing the best army, this heady mixture produced very nearly the worst. There was no sense of unity or combined purpose. The mercenaries only served together for money, and when matters grew difficult were prone to desert. The result was a motley crew, the product of all their national weaknesses rather than their strengths. Command was difficult with everyone speaking different languages, and national hatreds often rose to the surface making the task of Charles' officers unenviable.

At the battle of Grandson in 1476, the rashness of the Swiss should have handed victory on a plate to any calm and well-balanced opponent. One of the Swiss divisions – from Berne, Freiburg, Basle and Schwyz – had rushed off and made contact with the Burgundians, leaving the rest of the Swiss army behind. Victory was there for the taking. As the dense Swiss column came over the brow of a hill it found itself facing the entire Burgundian army, 18,000 strong. Charles ordered his cavalry to wipe out this solitary phalanx, but this was a task easier to order than to carry out. After the first wave of horsemen had been driven back, Charles led a second wave of lancers, but this also failed to penetrate the bristling mass of Swiss pikes. Undeterred, the Swiss now attacked the Burgundian footsoldiers in the centre. Imagining himself to be Hannibal reborn – and the grassy hills of Switzerland to be the broad plain of Cannae in Italy – Charles tried a double encirclement of the Swiss phalanx, withdrawing his centre and ordering his flanks to close round on what he assumed to be the doomed Swiss. Unfortunately Charles was not Hannibal that day, nor any other. Everything fell apart in his hands. Instead of moving like clockwork to achieve a stunning victory his soldiers behaved as if they had never seen each other before, had never seen Charles before – except on payday – and would be far happier fleeing and making for the nearest alehouse to spend their life savings.

As the Burgundian centre was 'withdrawn' by Charles, the rest of the army immediately assumed that it was in retreat, and fled. They were acting out of sheer panic, for the other divisions of the Swiss army had not even appeared on the field. The cause of the fiasco had been the lack of confidence that each man had in his fellow, and the fact that the different national groups had no experience of many of the finer points of Charles' tactics. Few generals of the time were as well versed in the military arts of the ancient world as Charles. Unfortunately, every time he tried something clever, it went wrong. There was no point in telling the enemy who beat you that you had been using the tactics of Julius Caesar and that by rights they should have lost the battle rather than won it.

While Charles' defeat at Grandson could be put down to poor discipline and panic, his reverse at Morat was a far more serious affair. Morale in his army – never good at the best of times – had slumped to rock bottom. He failed to appreciate that, however skilled they might be in the use of their weapons, men need to be confident in their commander and think that they have some chance of winning rather than being skewered on a long

Swiss infantrymen of the early 15th century equipped with their fearsome national weapon – the halberd. Swiss footsoldiers, fighting in massive phalanxes, inflicted a series of defeats on Charles the Bold of Burgundy in the 1470s.

Swiss pike. But Charles was not known as the 'Rash' for nothing. Within two months he was back and ready to take more punishment.

Charles now laid siege to the small town of Morat, which stood alongside the lake of the same name, and had a garrison of just 150 Bernese soldiers. He constructed a powerful defensive position, but he placed it in the most ridiculous place, with its back to the lake so that retreat in the event of a disaster was quite impossible. As if this was not bad enough he failed to make allowances for the fact that his front was only a short distance from a dense wood, through which an enemy army would be able to pass quite unnoticed. It was an object lesson in how not to select a military site.

The Burgundian army at Morat was even larger than at Grandson, numbering as many as 20,000 men, and Charles made it clear by his provocation of the Swiss that he had come intending to fight. Having selected what he thought to be a good position, he lined his palisade with gun emplacements – he was not intending to be caught in the open this time.

The Swiss assembled at Ulmitz on 21 June 1476, on the other side of the wood facing Charles' camp. There were perhaps 25,000 pikemen, as well as a number of feudal cavalry led by Réné of Lorraine. Surprisingly, Charles made no effort to send out scouts to locate the Swiss and they were allowed to move through the wood without any of the Burgundians seeing them. The first Charles knew of their whereabouts was when they came pouring out of the trees in front of his position.

As it happened, the day being wet and dark, he had sent many of his men back to camp and his palisades were defended by a skeleton force. As soon as the Swiss charged the Burgundian position, the defenders fled back towards their camp, colliding with the troops who were coming up to help them. Pandemonium broke loose. The Swiss simply drove the Burgundian infantry like cattle, many of them falling into the lake, while the rest fled in all directions. The losses in Charles' army were devastating. Although reports of a body count of 12,000 may be exaggerated, the encounter was undoubtedly more of a massacre than a battle.

Yet still Charles did not know he was beaten. With courage that was less admirable than laughable, he began gathering another army, though his native Burgundian troops were sadly reduced and few mercenaries wanted to follow him. But stirred into action by the surrender of his fortress of Nancy to Réné of Lorraine, Charles responded with typical rashness. Within days he had reached the city with a large army. This placed Réné in a difficult position, as he was hopelessly outnumbered, but he was

The death of Charles the Bold, last Valois duke of Burgundy, at the battle of Nancy, 5 January 1477. Under Charles the Burgundian state had reached the peak of its power, but he failed to make it a kingdom independent of France.

able to recruit over 8,000 Swiss pikemen as mercenaries and these men soon approached Nancy, where Charles was waiting for them. He chose another defensive position, this time one that could only be approached up a narrow path. He thought that this would break up the Swiss phalanxes and allow him to use his artillery to good effect. Drawing on the English tactics of Edward III, Charles flanked his position with archers and handgunners, and filled the centre with dismounted men-at-arms and knights. In front of all this he positioned his 30 large cannon. But as at Morat, he had selected a position from which there was no possibility of retreat, nor was the position very suitable for manoeuvring. The Burgundian army would have to stand its ground and conquer or die.

On 5 January 1477 the Swiss held a council of war. Having surveyed the Burgundian position they decided not to attack frontally, but to try to attack the flanks. A feint assault on the Burgundian centre was launched, but the main Swiss strength – some 7,000 foot and 2,000 horse – would strike Charles' right wing. The snow was thick on the ground as the Swiss flank-attack reached the Burgundian position. Taken unawares by this angle of attack, Charles ordered his artillery to be swung round but few of the guns could be brought to bear. Although the Burgundian cavalry bravely drove off the Lorraine horse, they could do nothing with the bristling pikes of the Swiss. In desperation Charles ordered his archers to concentrate on the Swiss phalanx as it drew inexorably closer, but it was all too late. With irresistible force the Swiss stove in the Burgundian position and a general mêlée ensued, in which 7,000 Burgundian foot were slaughtered. Fighting in this last hopeless battle, Charles was struck by a Swiss halberd and his head was cloven from top to chin. After the battle his body was recovered from a frozen pool, where it had been partially gnawed by wolves. It was a heroic end for a courageous but hopelessly inept commander – a pretender to the ranks of the great captains.

THE BATTLE OF DREUX (1562)

What do the two worst commanders of the age do when they meet in battle? Answer: capture each other. The battle of Dreux is unique in the annals of military history for being the only battle in which both army commanders got themselves captured at the same time. To compound the farce the French Catholics also had their second-in-command captured as well. It was that kind of battle.

The extended civil war between Catholics and Protestants in France in the 16th century – known as the Wars of Religion and in fact consisting of eight different conflicts – was marked by numerous battles but very little military skill, at least until the intervention of the Spanish Duke of Parma in the 1580s. In some ways it resembled the English Civil War, with enthusiastic amateurs and incompetent professionals committing blunder after blunder and generally getting away with it. The Constable of France – Anne de Montmorency – should by rights have been his country's foremost soldier. He probably thought he was. But during the Italian Wars of the earlier part of the 16th century he had earned a reputation as a *beau sabreur* and had come to believe that there was nothing more to learn about the art of command. As long as one could shout 'Charge' and follow the horse's nose one could lay claim to being a general. As one of France's leading Catholics, de Montmorency was certain to command their armies when war came. Age had not mellowed him, and at 74 he was as cantankerous as any man in France. Although there were far more able officers – the powerful Duke of Guise for one – they had to play second fiddle to Montmorency.

On the Protestant side command went by rank to the Prince of Condé, probably the only general in France capable of matching Montmorency blunder for blunder. Alongside Condé fought the Admiral of France – not a sailor but a soldier – the noble Gaspard de Coligny, a far better military thinker than Condé, but forced, like everybody else, to concede precedence to a prince of the blood.

On 19 December 1562 the Protestant army under Condé and Coligny was advancing along the road towards Dreux, to the west of Paris. Their casual approach to warfare was clearly evident by the fact that although they had with them nearly 4,000 cavalry, as well as 9,000 infantry, they had not bothered to send out scouts to discover the whereabouts of any Catholic forces. Condé felt inclined to leave petty matters of that kind to some junior officer or another. It was not something for princes of the blood to concern themselves with.

Condé was stopped in his tracks by the sudden

spectacle of the Catholic army, drawn up in front of him, blocking the road. Montmorency, dredging up from the dark recesses of his mind the need for reconnaissance, had scored the first point in the stately game of tactics against the Prince of Condé. Condé was piqued. Montmorency had picked a good position, flanked by woods on both sides.

Ironically, France's best general, the Duke of Guise, had refused a senior command in the Catholic army on the grounds that he did not want to bear responsibility for a defeat which was almost certain with a man like Montmorency in charge. As a result Montmorency and St André shared command, with Guise acting in only a very minor capacity.

Coligny was impressed by the position that the Catholics had taken up and was convinced that Montmorency, having done so well, would not want to risk too much by attacking. He therefore advised Condé simply to by-pass the enemy and engage on another day when the situation might be better. But Condé, once roused, was difficult to control. He claimed that the armies were too close to disengage and that a battle was inevitable. The Huguenots therefore deployed for battle, with Coligny's wing facing Montmorency and Condé's facing St André.

In the early afternoon Condé, despairing of the Constable ever leaving his defensive position, began to move his men westward as a prelude to disengaging. No sooner had the Protestants turned to march, however, than Montmorency – seeing an exposed flank – decided to attack it. The decision was to be his undoing. As he charged forward his cavalry were swept away by Coligny's horsemen. Montmorency himself, fighting in the front line, was knocked from his horse, slightly wounded and taken prisoner. Guise, far back in the Catholic ranks, breathed a sigh of relief. With the old man gone they might stand a chance. For the time being, however, Coligny's wing had won a complete victory. Fleeing Catholics carried news of complete disaster to the Queen Dowager, Catherine de Medici, who observed dryly, 'We shall have to learn to say our prayers in French.'

But on the left things had gone rather differently. Condé found himself facing not just St. André's cavalry but a formidable phalanx of Swiss pikemen. As if bent on self-destruction Condé, riding at the front of his cavalry, selected the Swiss as suitable opponents and charged at them hell for leather. Forgetting that there were any other troops on the battlefield, he concentrated his whole attention on

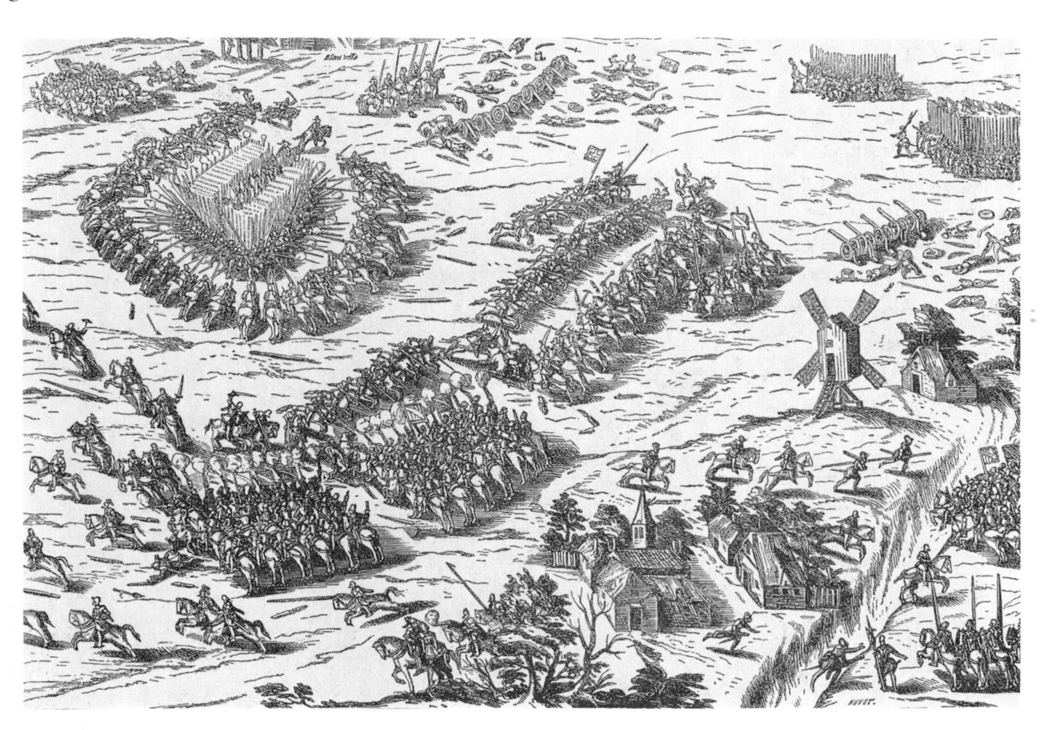

The battle of Dreux, 19 December 1562, fought between Huguenots and Catholics during the French Wars of Religion. In the confusion of a hard-fought battle, both the opposing commanders, Montmorency and Condé, were captured.

ridding the earth of this particular group of mercenary ruffians. Time after time he threw himself into the mass of pikes. But the Swiss line could not be broken and as the Huguenots recoiled from their pikes, Guise advised St. André to order a general attack. It was the one sensible decision in the whole fight. Condé, seeing the day turn against him, gathered a few horsemen and charged into a regiment of Catholic *Gensdarmes*. He was immediately unhorsed and taken prisoner.

Hearing – too late – of the prince's discomfiture, the victorious admiral gathered what cavalry he could and charged across the field to try to rescue the right wing. He arrived just in time to rout St. André's horsemen and to take the Marshal prisoner. Guise, meanwhile, was forced to assume command of the army. However, Coligny could not overcome the Catholic infantry and was forced to withdraw from the field, leaving the honours of the day even.

After the battle the Queen Dowager, Catherine de Medici, persuaded the two great prisoners, Montmorency and Condé, to negotiate an end to the fighting, and consequently an exchange of prisoners. Neither of the pair was to benefit much from his experiences. The standard of leadership in the wars did not improve until the advent of the Huguenot leader, Henri of Navarre, in the late 1570s.

THE BATTLE OF KERESTES (1596)

The ability to exploit success is as important a quality in a commander as the ability to gain victory in the first place. During the Danubian Wars between the Austrian Habsburgs and the Ottoman Turks, the failure of the Archduke Maximilian to control his successful troops led to an astounding reversal of fortunes.

A series of military setbacks in 1596 convinced the Turkish Grand Vizier, Ibrahim Pasha, that there was no alternative but for the sultan himself – Mohammed III – to lead the forces of the Ottoman Empire in a campaign on the Danube. The Ottoman army – numbering as many as 100,000 combatants – advanced on the city of Erlau, which surrendered to the Turks through the treachery of its garrison of mercenaries. The Christian commander, the Habsburg Archduke Maximilian, was joined by Prince Sigismund Bathori of Transylvania, and the combined force marched towards Erlau to bring the Turks to battle. Maximilian's army numbered some 40,000 men with 97 cannon, and thanks to the Transylvanian reinforcements it was now strong in cavalry as well. Twelve miles south of Erlau, near the town of Kerestes, Maximilian found the terrain to his liking and prepared a defensive position behind a marshy depression.

On 23 October the main Turkish army appeared in front of the Christian defences. However, it declined to attack straightaway and set up camp no more than a mile away. The following day the Turks, commanded by the Pasha of Buda, launched a full-scale frontal assault, but were driven back with heavy losses. It was now that Maximilian's main problem – namely that his army was composed of many different national groupings – began to surface. As supreme commander he had under him a series of able if often impetuous generals, such as the Hungarian Palffy, the Austrian Tieffenbach and the German Schwarzenburg. Keeping these men under control was quite a task.

Having suffered a severe reverse Mohammed was keen to return to the delights of his harem. Only with some difficulty did Ibrahim persuade him that his departure would result in the complete collapse of Turkish morale. Eventually, Mohammed agreed to have one more try and on 26 October, before dawn, the Turks renewed the attack. They swept down on the Christian position in a great half-moon formation, with the élite Janissaries in the centre drawing the artillery, which was chained together. Unfortunately for Mohammed, Maximilian had read his opponent well and was waiting for just such an attack, having kept his army at battle stations since before dawn. The Hungarian and German cavalry charged out to meet the Turks and soon Palffy and Schwarzenburg had routed their opponents, who promptly fled, not stopping until they had travelled some 20 miles. Anticipating a problem, the Archduke had emphasized the need for commanders to keep their men in check and not to pursue the beaten foe. But nobody took any notice of him and the Christian cavalry simply charged after the Turks, with their commanders urging them on rather than calling them back. Palffy was insistent that Maximilian should not spoil a half-won victory, and he was supported by others, including Prince Sigismund. Maximilian bowed to their entreaties, though it would hardly have mattered had he refused, for already the Christian soldiers were descending on the Turkish camp like a swarm of locusts. A general sack of the camp followed, with the sultan's mutes and eunuchs being slaughtered and his gorgeous tent torn into shreds by rampaging Christian troops.

So intent were the Germans and Hungarians on plundering the Turkish camp that they completely failed to notice a new Turkish force approaching their flank. It was a squadron of Turkish cavalry commanded by Cicala, a Christian renegade. Cicala had avoided the general rout of the sultan's army and had waited until the Christians dropped their guard before sweeping into the camp and taking the plunderers entirely by surprise. Panic broke out among the Christians and soon thousands of them, cavalry and foot, were running back across the plain towards the Imperial lines, where Maximilian had just two regiments of horseguards left to protect him. It was an extraordinary turnaround. Soon the whole Habsburg army was in retreat, pursued by Cicala's Turks. It is estimated that 6,000 Christians died in the pursuit alone.

When he heard the miraculous news Mohammed could only see the will of Allah working through Cicala. Disgracing the failed Grand Vizier Ibrahim, Mohammed elevated the opportunistic renegade in his stead. Although the Turkish losses had been very heavy in the earlier fighting, victory had been snatched from the jaws of defeat. Yet, however brilliant had been the intervention of Cicala, the Christians must look back on Kerestes as a victory thrown away through the collapse of discipline by both soldiers and commanders at a crucial moment. It is perhaps harsh to blame the Archduke for the fiasco, but he should have imposed his authority more completely on his subordinate commanders.

BUCKINGHAM'S EXPEDITION TO THE ILE DE RÉ (1627)

The financial troubles of Charles I, which eventually led to his open breach with Parliament in 1629, owed much to the military incompetence of his friend and favourite, the Lord Admiral, George Villiers, first Duke of Buckingham. In just four years, from 1625 to 1628, Buckingham involved England in four ruinously expensive military ventures in Europe. Pressing hard upon the heels of the ill-fated Cadiz expedition of 1625 came a plan to go to the aid of the French Huguenots at La Rochelle by capturing the nearby island of Ré. As usual, Buckingham's ambitions exceeded his capacity to realize them. When he pressed Lord Treasurer Sir John Coke to find yet more money to finance ships and men, he was told, 'the inferior orbs of action are ready enough for motion. . . . Money, the *primus motor*, doth retard us all in our ends.' Buckingham would have done well to take heed of this wise assessment. Instead, he went ahead with his plans, promising to pay for everything himself, and relying on the king's capacity to raise loans from an unwilling population. The result was that the expedition set off short of every essential.

Things had not changed much in the brief period since Wimbledon's ships had limped home from Cadiz (see *The Guinness Book of Military Blunders*, pp. 142–5). The fleet still comprised leaky survivors from the reign of Elizabeth I, manned by ragged and underfed crewmen and commanded by foppish officers who knew little of the sea. Sir James Bagg – 'bottomless Bagg' – who had provisioned the Cadiz expedition with mouldy food and infected water, had kept the Duke's favour and was now given the task of equipping the new force. And the English troops, who had given such poor service under the command of Viscount Wimbledon, were of the same poor quality as those who had been levied in 1625. The Hampshire levies were described by one writer as 'such creatures as I am ashamed to describe them'. From inland counties came 'poor rogues and beggars' without proper clothing and many in poor physical shape. Buckingham was shocked to hear this – which shows he could not have been listening when Wimbledon complained of the same problem just two years before. He ordered the Lords Lieutenant of the shires 'to take more care to send young and able-bodied men, well-clothed and fit for service'. He might have saved his breath, for none of them took any notice. Foreign campaigns were an opportunity to get rid of vagrants and undesirables. Nobody in their right mind would send the best men to die abroad.

With money in such short supply that soldiers and sailors remained unpaid throughout, Bucking-

George Villiers, first duke of Buckingham, in a sketch by Peter Paul Rubens. As Lord Admiral of England, Buckingham personally led the expedition to relieve the Huguenots at La Rochelle in 1627. He perished the following year, at the hand of an embittered veteran of his own disastrous campaigns.

ham found he had to make economies in his own preparations: just a few thousand pounds were spent on his linens, silks and his favourite gold and silver buttons. Buckingham had decided to make up for his lack of military experience by at least looking the part. As one observer wrote, 'His serious intention is also shown by the military costume which he wears, with an immense collar and magnificent plume of feathers in his hat.' Buckingham also took the following aboard the flagship for his personal use: £50 worth of books for quiet reading in between battles; a musician with a harp to 'soothe the savage breast'; a farmyard full of animals for his personal sustenance, including oxen, milch cows, goats and chickens; a magnificent coach and horses, as well as footmen, pages and coachmen; satin and velvet suits for evening wear after a hard day in the trenches; and an ornamental icon of the French king's wife, Anne of Austria, on whom Buckingham had developed a crush – which was set up on an altar aboard the flagship and at which he worshipped daily. In all Buckingham had restricted his personal spending to £10,000 – enough to have equipped and maintained a whole regiment of soldiers for six months.

The army that finally assembled at Portsmouth consisted of seven regiments of a thousand men each, led by some experienced officers. These included Sir Edward Conway, who had endured the Cadiz expedition in 1625 and yet still came back for more. Buckingham had offered a regiment to the Earl of Essex, who had also been to Cadiz, but the Earl wisely refused. The general of the army, under Buckingham's overall authority, was to be Sir John Burroughs, a third survivor from Cadiz.

The fleet sailed out of Portsmouth harbour to the cheers of crowds of onlookers and the strains of musical instruments. It was a moving scene, played out a hundred times in English history from the time of Edward III right down to the Falklands War of 1982. And yet Buckingham's voyage had one slight difference. The gracious people of Portsmouth roundly booed him and chanted a scurrilous poem which said,

> Most graceless Duke, we thank thy charity,
> Wishing the fleet such speed as to lose thee,
> And we shall think't a happy victory.

Whatever Buckingham thought of this poem, he was secure in the confidence of the only person who mattered to a king's favourite – the king himself.

Buckingham sailed in the *Triumph* along with over a hundred ships, 40 of them supplied by the king. However splendid a sight it was to the crowds on shore, the fleet's power was more apparent than real. When they chanced upon some pirates in the English Channel and tried to give chase, the sailing deficiencies of the elderly English warships were soon revealed. Nevertheless, Buckingham's confidence was high as he anchored off La Rochelle on 10 July. He sent Sir William Beecher ashore to tell the inhabitants of the city that the Duke of Buckingham had come to rescue them from the French king. The Rochellois told Beecher they were too busy to talk to him as they were having a religious fast, whereupon Beecher returned to Buckingham in a huff. The Huguenots, it seemed, were not impressed by the news of his arrival. Buckingham was baffled but decided to go on with his plan, which was to land on the island of Ré, giving him control of the approaches to La Rochelle. To spy out the land he sent a ship's boy to swim ashore from the flagship, with instructions to run a mile inland to see if there were any French soldiers about. The boy was soon spotted and pursued back to the shore by a group of Frenchmen, but he managed to get back to the *Triumph* in one piece and reported that the area Buckingham had selected for his landing was unfortified. The admiral therefore instructed his officers to prepare landing parties. However, by now news that there were naked English boys playing on the beach had brought out the locals. From the top of the *Triumph*'s mainmast Buckingham could clearly see bodies of French infantry drawn up on the beach. He would have to fight his way ashore.

As the landing parties were rowed towards the beach the English warships opened fire to keep the Frenchmen from advancing into the surf. The first men ashore were from Sir John Burroughs' and Sir Alexander Brett's regiments, and with them went Buckingham himself. Having been cooped up in the holds of their ships for so long the English soldiers were more in the mood for some gentle leg-stretching than a stern fight. As they waded through the sea, tipped as it was with white Biscay surf, they lingered, playfully splashing each other and paddling in the shallows. Others threw themselves gratefully down on the white sand to rest in the sun. It would have been an idyllic scene had not the French insisted on trying to attack them with 15-foot pikes. It was only when Buckingham took up a cudgel and began driving the men ashore that the English took up their weapons and got into some sort of order. A sharp fight now took place, with the English succeeding in driving the French off the sand. However, before they could reform, a force of French cavalry charged down the beach and drove the English footsoldiers back into the sea. The holiday was over. The English infantry – about 1,000 strong – now formed up into a solid phalanx of pikes which the French horsemen could not penetrate. Protected by the pikemen, English musketeers shot the French down in droves, forcing them to retreat. Meanwhile, the French infantry – some 1,500 strong – had reformed and advanced on the English foot. To their misfortune, the French found that their pikes were shorter than the Englishmen's and, after a good deal of pushing and shoving, they abandoned their pikes and starting throwing stones instead. What had begun as a bloody demonstration of the art of 17th-century warfare ended in a kind of playground scrap. But all in all it had been a bloody struggle. Many French notables had been slain and, although no English soldiers had been killed, as many as fourteen high-ranking English officers had been lost, including two of Buckingham's senior colonels. Nevertheless, Buckingham had demonstrated not only his own personal courage, but his capacity to lead and motivate his motley army.

The French commander, Marshal Toiras, withdrew his troops to the town of St. Martin, leaving the English masters of the beach. But Buckingham now showed his inexperience. Instead of pursuing the beaten enemy right into the citadel of St. Martin he told his men to dig in on the beach in case the French made another attack. It was a bad mistake, for prompt action might have made him master of the whole island that day. Buckingham, however, was in a pensive mood and may have felt the need to rest, bind up his wounds and bury the dead. He was young and still in love with chivalry, so when three French noblemen, hurt in the fighting, asked his permission to go to the mainland to have their wounds dressed, he sent his own scarlet-lined barge to carry them over the foam, even lending his musicians to cheer them or soothe their suffering.

Buckingham was becoming disenchanted with the Rochellois. He had been led to believe that – like a damsel in a tower – they lived only to hear the sound of his horse's hooves coming to their rescue. In fact, they cared not a jot for his mission or for the fact that English soldiers were dying on their account on the beach at Ré. The local people, it appeared, were far more concerned with getting in the harvest. Meanwhile Buckingham had fallen out with Sir John Burroughs. The latter, a dour professional soldier, found the Duke's chivalrous attitude totally ridiculous. He was in favour of getting on with the business of taking the island and making mincemeat of the French. Burroughs persuaded Buckingham to move against St. Martin and the town was soon in English hands, but Toiras and his troops withdrew into the powerful citadel and defied Buckingham to prise them out. In spite of his defiance Toiras must have known that his position was parlous indeed. Short of water and provisions, he could not expect to withstand a prolonged siege without help from the mainland and this the English fleet would surely intercept. Buckingham would have been wise to leave the garrison in St. Martin and move his troops onto the mainland, for ostensibly the island was his already. But he considered it a dishonour to leave the island without taking its main stronghold, and so he committed his troops to a lengthy and unnecessary siege operation, for which his men were ill-prepared.

If Toiras had his difficulties inside St. Martin, they were as nothing compared to those problems that faced Buckingham on the other side of the battlements. In the first place, the duke was short of money to pay his men, and in the second he needed the 4,000 reinforcements that he had been promised when he left England. But how was he to pay the first and finance the second? It always came down to money – and he never had enough of it.

Buckingham's master gunner – a swaggering rogue – assured the Duke that as soon as his guns were set up he would batter the citadel into surrender. Instead, bafflingly, he opened fire on some nearby windmills. This infuriated the French and they opened a devastating counter fire which silenced the English guns in a matter of minutes. Clearly the master gunner did not know his trade. Nor did the engineers who were digging the entrenchments. They dug them so far away from the citadel that the musketeers were out of gunshot range. For nearly six weeks the soldiers occupied the trenches in complete safety but quite unable to inflict any harm on the garrison. This amused the French so much that when a decision was made to dig new trenches somewhat closer to the fort the French, in the words of one of the English officers, 'cheeringly told us that they thought we had been lost and wondered where we had lay hidden all the while'. As usual the so-called English specialists were of lamentable quality. It transpired that the 'chief' engineer – who was supposed to know everything about forts – had only been a labourer before and he had the trenches dug with the wrong side fortified. Buckingham was so frustrated that he eventually appointed his own gardener to design the trenches.

It was a very wet summer and the English soldiers had to spend long periods in the trenches, eating poor food, and succumbing to disease. Buckingham was forced to supplement his soldiers with 500 men from the fleet, but even so he was still short of men.

Above all, Buckingham needed to prevent the French from getting supplies into Fort St. Martin from the mainland. He therefore asked his 'engineers' to devise methods of blockading the port entrance. One method tried was a floating stockade of masts and timber, chained together, but it was

Cardinal Richelieu, chief minister of the French King Louis XIII. Richelieu's attempt to capture the French Protestant stronghold of La Rochelle was the catalyst for Buckingham's expedition to the Ile de Ré.

soon smashed to pieces by the waves. The next effort was a kind of pontoon, made by linking ships together. Buckingham was so pleased that he boasted that 'not a bird' could get into St. Martin. He was wrong. The stormy seas broke up his pontoon and it was blown out to sea where it veered about in a crazy fashion, threatening the English ships and causing some to lose their anchors. A third attempt was a floating 'island', constructed of upturned boats, under cover of which the English soldiers could shelter as they drove off any relief vessels from the mainland. Again the wind was unkind, lifting it up and throwing it down so heavily that it began to break up, and the waves completed its destruction.

Buckingham now received a series of most discouraging letters from England. All his friends – who had wished him well, watched him go and prayed for his success – wrote to tell him that he had no chance of getting any money to continue the expedition. In desperation Buckingham wrote to his mother to ask her for help. She wrote back that she would have liked to help but she had just bought a new property and therefore had a cash-flow problem. Buckingham's wife, who had also received one of his begging letters, sent him £200 of her housekeeping money and reminded him that their roof needed mending. When the king wrote to tell Buckingham that he too lacked the funds to help it was a virtual death sentence on the whole operation.

Meanwhile, in England, reinforcements for Ré were being assembled at Portsmouth. Surprisingly 400 men arrived on time, but there were no transports ready to take them to France. Weeks passed and the transports remained in the Thames, unable to sail round to Portsmouth because the munitions they were supposed to carry had not been released by the ordnance office. Tied up as well was the £14,000 that Buckingham needed to keep his army in the field. Eventually he received just £10,000 and had to whistle for the rest.

Buckingham, meanwhile, had not lowered himself to the level of the men who squabbled with him over pay. When he heard that the French governor Toiras had inquired if there were any melons on the island he generously sent him a dozen of the finest melons. Responding in kind, Toiras sent Buckingham half a dozen bottles of orange flower water and some boxes of Cyprus powder, which the duke sent back to England for his wife. She threw them away, however, fearing that they were poisoned. It sometimes seemed to the English soldiers, living in miserable conditions, that Buckingham and Toiras were more like friends than enemies. Nevertheless, there was a harsher side to the duke, which revealed itself in his sending all the Catholic women and children of the island into St. Martin to increase pressure on the citadel's limited food supplies. The belated arrival of reinforcements for Buckingham in the form of 2,000 men from Ireland further depressed Toiras' garrison, who could see that the English had no intention of leaving the island as they had hoped. Toiras now called for three volunteers to swim to the mainland and fetch immediate help. Only one of the three survived the swim but he was able to convey news of the parlous state of the St. Martin garrison. Relief ships would be needed to break the English blockade.

Meanwhile, Toiras sent an officer to Buckingham to discuss terms of surrender. Buckingham at once showed himself to be a poor negotiator. Having praised the garrison for its gallantry – this was strictly unnecessary but good for Buckingham's image – he asked the Frenchman to come back the next day, which gave the French another 24 hours of respite. When the Frenchman returned the following day, Buckingham – astonishingly – suggested to him that Toiras should draw up his own terms of surrender. After a three-hour delay Toiras agreed but said he would need another 24 hours, to which the duke agreed. Buckingham had blundered. Though he could not have known it, the French relief fleet was preparing to sail that very night.

The English sailors were half asleep when the small French pinnaces sailed right through the blockading fleet. They had reached St. Martin and had begun to unload their precious cargoes before the English were fully aware of what had happened – by which time it was too late. Fire ships were sent into the harbour to burn the French vessels but to no avail. When dawn came the French garrison were seen on the walls jeering at the English and parading whole hams and turkeys on their pikes.

This was a great blow to English morale. Starvation had been Buckingham's main weapon against St. Martin and that had clearly failed. The army commanders now pressed the duke to pull out of Ré as there was no point in carrying on the siege. As they said:

> By the late coming of their succours and supplies they were kept in such continual weakness that they could not advance their works . . . That by extreme duties, and the immoderate eating of grapes, the soldiers were so wasted as there do not now remain above five thousand able men and two thousand five hundred sick men, and the disease runs on so violently as worse is daily to be feared.

Buckingham hated to abandon the island, but even he realized that he had no choice. Disease was sweeping his camp and the French were preparing an expedition from the mainland. If he did not leave soon he would never get away. Yet he decided to have one last try at storming the citadel.

On 27 October, to the sound of the English singing hymns, the assault began. Men who only days before had gone 'bare-arsed' from dysentery now – with apologies to Shakespeare – 'stiffened their sinews' and 'stood like greyhounds in the slips'. It is said some even 'lent the eye a terrible aspect'.

Then they drove the French from their outworks and came up to the walls of the citadel itself. Scaling ladders were brought forward, but they were too short. The first men up found their noses pressed against the stone walls, while the French above them dropped logs and heavy weights on their heads. For two hours the English milled about like sheep at an auction while the French shot them down at their leisure. Buckingham must bear much of the blame for the fiasco. After all, once he had reached the decision to leave the island there was no point in the final attack – and someone should at least have made sure that the ladders were long enough.

Buckingham now ordered an evacuation of the island, and only just in time. Marshal Schomberg had arrived on Ré with 6,000 reinforcements and soon the English would be cut off from their ships and annihilated. The retreat was to be made by way of a causeway connecting the Ile de Ré to the smaller Ile de Loix, where the English fleet would be waiting. Buckingham had ordered his engineers to build a redoubt at the end of the causeway where troops could assemble to drive off the French attacks. But the engineers had built the redoubt at the wrong end, allowing the French to attack the English as they tried to pass down the causeway, which was 500 feet long but just 4 feet wide. The engineers had also failed to put rails along the causeway to prevent the men falling into the salt marshes on either side. It was a costly blunder. As soon as the English troops reached the causeway they started pushing and jostling to get away first and pandemonium ensued. Dozens of men fell into the marsh and were drowned. French troops now poured towards them and heavy fighting engulfed the access point to the causeway. Hundreds of men were pushed into the marshes by the French pikemen. Once all the English had reached the Ile de Loix a major battle took place in which the desperate English succeeded in driving off their assailants and chasing them back down the causeway. Not entirely surprisingly, Buckingham did not order a pursuit and the grim embarkation took place. The English were leaving Ré with 5,000 fewer men than when they had arrived; disease had caused most of the losses but battle casualties had been high. As the English fleet sailed away from La Rochelle they encountered ships coming from England with supplies and reinforcements.

The failed Ré operation rivalled Wimbledon's expedition to Cadiz in terms of incompetent handling and poor planning. So hated was Buckingham that it is said that his return caused more sorrow in England than the loss of all the other men had done. Once again he was pilloried in popular doggerel:

> And art returned again with all thy faults,
> Thou great commander of the all-go-naughts,
> And left the isle behind thee? What's the matter?
> Did winter make thy teeth begin to chatter?

THE BATTLE OF MALPLAQUET (1709)

The battle of Malplaquet occupies a special place in military history, being almost certainly the most 'pyrrhic' of all 'pyrrhic victories'. It also represented to contemporaries the kind of bloody and pointless fighting that is symbolized today by bloodbaths such as the Somme in 1916 and Passchendaele in 1917. Malplaquet was, in fact, the bloodiest battle of the 18th century, and was not to be approached in butchery until Borodino in 1812. But the Allied commanders at Malplaquet – John Churchill, duke of Marlborough and Prince Eugène of Savoy – were a world away in ability from First World War dunderheads like Haig, Rawlinson and Gough, which just goes to show that butchers are not always blunderers.

Malplaquet was not only a very bloody battle, it was an unnecessary one as well. During the Allied siege of Mons in 1709, Marlborough and Eugène – with 100,000 men – found themselves confronted by a large French army of 90,000 under Marshals Villars and Boufflers. It was a last desperate effort by the French to save themselves from complete defeat in the war. The Allies had no reason to fight the French, particularly as Villars had established a strong defensive position and could dictate the way the engagement was conducted. It was unwise of Marlborough to allow himself to be led into a battle that he did not need to fight. The French strategic position would not be much worsened by defeat, but victory or even a drawn battle would work wonders to restore French morale and prestige.

Commanding an army of 100,000 – the biggest ever assembled in European history – was a task beyond the powers of even the most skilled generals

John Churchill, first duke of Marlborough, on the morning of the battle of Malplaquet, 11 September 1709. The battle – the last of Marlborough's great victories – was marred by heavy Allied casualties.

like Eugène and Marlborough. In defence the matter might have been simpler but in attack, and Marlborough as always intended to take the offensive against the strong French positions, a commander needed to lean heavily on his subordinate commanders. And at Malplaquet Marlborough was to find himself ill-served by the heroic but idiotic Prince of Orange.

According to his rank and the number of troops he contributed to the alliance, the Prince of Orange could hardly be denied a senior position in the army, and when Marlborough gave him command of the left wing, facing Marshal Boufflers in the Bois de Lainière, he had every expectation that the more experienced officers around him would be able to control the young man's youthful ardour. Yet the left wing was facing a more difficult task than even Marlborough realized. Boufflers had skilfully positioned 20 cannon in the wood so that they would enfilade the Allied cavalry as they advanced. For once the duke's acute observation had failed him.

The overall plan of action was for the Allied right wing under Schulenberg to begin the assault on the French positions, preceded by an artillery bombardment. Thirty minutes later it was the turn of the left to advance – but not in a 'hell for leather', 'death or glory' way. The role of the Prince of Orange was 'containment' rather than penetration. Whether Orange understood this or not is open to question. He may have been piqued at being allocated such a small part in the action or he may have been simply obtuse, which is more likely. In any event, he did not carry out Marlborough's orders. He waited 30 minutes to allow Schulenberg to develop his attack on the right, but once the time was up, he summoned his 30 Dutch and Scottish squadrons of horse and charged full tilt towards the French positions of De Guiche and D'Artagnan. It was like the Charge of the Light Brigade but on an immeasurably larger scale. Thousands of horsemen rode full tilt into a hail of gunfire from front and right flank. Twenty or thirty cannon poured cannister and grape shot into the Dutch ranks, slaughtering men and horses in a dreadful carnage. General Oxenstiern fell early and within 30 minutes of going into action the Prince of Orange, who had two horses killed under him, had suffered over 5,000 casualties. Time after time his horsemen rode into the guns, until famous regiments like the Dutch Blue Guards had been almost annihilated. When a second attack was launched two more generals – Spaar and Hamilton – were blown from their horses and killed, yet the Prince himself seemed to bear a charmed life.

Meanwhile, on the right, after hard fighting Schulenberg had broken into the French defences, but at great cost. However, as he surveyed his hard-won gains, Marlborough received a desperate message telling him to come at once to the left wing where a terrible reverse had been suffered. The messenger was the Dutch Deputy Sicco van Goslinga, who had been shocked by his own prince's behaviour. Both Marlborough and Eugène rode over to the Prince of Orange's headquarters, just in time to prevent the headstrong young man from launching a third attack by his ravaged cavalry. It has been suggested that Orange was intent on winning some of the glory for himself by breaking the French lines, rather than carrying out his orders from Marlborough. At this distance one can only guess at his motives, yet had the duke not arrived in time to stop another charge it is possible that the entire left wing of the Allied army could have been beaten and driven from the field. Afterwards, Marlborough expressed himself content with Orange's performance. Under the circumstances, this can hardly have been true. The Dutch had suffered more heavily than any of the allies in the battle, with 8,600 casualties or 43 per cent of their men involved. It is said that the Dutch state carried the burden of this battle for years to come, with husbands, fathers and sons lost from every important family in the Netherlands.

Although the French were finally forced to withdraw from their positions, they had inflicted 25,000 casualties on Marlborough's army at a cost to themselves of some 12,000. Malplaquet was a tragic 'victory' for Marlborough which won him little. It did not even provide him with the peace he wanted. The war dragged on for another four years, though Marlborough never commanded in another battle. As one of his generals, Lord Orkney, wrote after Malplaquet, 'I hope to God it may be the last battle I see.'

A 'wanted' poster for Charles Edward Stuart issued during the Jacobite rebellion of 1745, offering a reward of £30,000 for his capture. The British government's first attempt to resist the Jacobites resulted in a humiliating retreat from Coltbridge.

THE 'CANTER OF COLTBRIGG' (1745)

The outbreak of the Jacobite Rebellion in 1745 found Britain virtually defenceless, with her best troops fighting on the Continent. Even the regular troops that were available seemed to prefer running to fighting, while their officers were either incompetent 'fops' or brutal martinets.

On 16 September 1745, Charles Edward Stuart and his small army were just eight miles from Edinburgh. Panic had broken out at the news and the only government troops in the vicinity of the Scottish capital – two regiments of ill-trained recruits, the 13th and 14th Dragoons, of whom not too much was expected – were quickly brought into action. The dragoons and the Edinburgh town guard took up positions at Coltbridge, watching the road from Stirling. However, the dragoons were feeling jumpy, and when a few Jacobite horsemen were spotted in the distance by their pickets – and a few shots were fired – it was as if the devil himself had suddenly appeared. First the pickets rode hell for leather towards Coltbridge. Then, as soon as they came into sight, the two regiments of dragoons panicked, mounted their horses and, with their officers in forlorn pursuit, fled south, many not stopping until they reached Prestonpans.

Even here they did not feel safe. As darkness fell one of the dragoons, out looking for forage, fell into a disused coalpit full of water and started yelling for help. At once the cry went up that the Highlanders were on them and the dragoons remounted their horses and fled again, this time as far as Dunbar. It was quite a ride. But every mile they covered enhanced the legend of Jacobite invincibility. Through village after village the English soldiers fled from their own fears, crying out that they had been betrayed and that the Highlanders were coming. Soon all the lowland Scots were changing their allegiance and, thanks to the fleeing dragoons, Charles Stuart was able to enter Edinburgh unhindered, where he took up residence at Holyrood Palace. From all over Scotland men flocked to join him. Rarely has such a ludicrous incident had such significant results.

THE BATTLE OF THE WABASH (1791)

In the early days of the new American Republic, making war on the Indians was far harder than it sounded. In the first place the Indians had an intimate knowledge of the virgin forest in which they lived, and in the second the troops that were sent to deal with them were the scum of the earth and took to their heels at the first opportunity. As a result the government of George Washington soon had two prime military blunders on its hands. First a professional blunderer – Brigadier-General Josiah Harmar – was ambushed in the forest after he had divided his ramshackle force into more separate parts than he could rightly remember. Then Arthur St. Clair, an ex-British regular with no experience of Indian fighting, crippled with rheumatism and gout, was chosen to lead an expedition in a harsh North American winter to avenge Harmar. It was a forlorn hope.

The quality of St. Clair's troops was no better than Harmar's, and disaster was inevitable from the start. The army that assembled at Cincinnati comprised shiftless Irish immigrants, jailbirds and public debtors. To make up the numbers Secretary for War Knox had, in desperation, called out the militia and to speed up St. Clair's departure everything was rushed and corners cut in providing food, munitions, clothing and transport. It did not take the militiamen long to make their presence felt. They ignored St. Clair's authority, refused army discipline, deserted when the fancy took them, ran away when danger threatened and, when reprimanded, tried to plunder the supply column. Yet one has to have some sympathy with their point of view. Most of them were 'badly clothed, badly paid and badly fed', and had short rations, defective powder and inadequate clothing for an autumn and winter campaign. Their muskets and small arms were broken or faulty, often lacking touch-holes. Their artillery carriages were rickety and sometimes lacked wheels. Their horses were usually short of fodder and, within weeks of leaving Cincinnati, frost and snow was killing the edible grass along the forest paths so that the horses starved.

The force that St. Clair led out of Cincinnati consisted of a mere 600 regular infantry, 800 enlisted men – the 'off-scourings of large towns and cities, enervated by idleness, debaucheries and every species of vice . . .' – and 600 militiamen. They had had to leave the fort earlier than planned to remove the new recruits from the temptations of the grog shops. The march began through heavily timbered virgin forest and a road had to be cut by woodsmen who walked ahead of the soldiers, thus slowing everything down to a crawl. Bad weather was beginning to close in and the cold nights of October soon showed the men that their clothes and tents would offer scant protection when the snows came. Many began to desert, and groups of hostile Indians were now sighted for the first time.

A few weeks out from Cincinnati St. Clair was overtaken by reinforcements – a rabble of 'debauched and idle-looking men'. In his diary Adjutant-General Sargent refers to these men as the merest 'trash'. Nor was the military efficiency of the 'army' improved by the camp-followers, a motley collection of women, children and tinkers. Some of the women were wives or mistresses of the soldiers and helped the men as washer-women, others were dissolute creatures who fell ill in the harsh conditions and died unnoticed on the march.

Periodically St. Clair tried to impose some discipline on his men. But in dealing with such poor material – the human flotsam of a frontier society – he found that his floggings only made the men desert in greater numbers. Officers were court-martialled for minor offences and on one occasion the whole army was halted while a gallows was built and three men executed. Such incidents were hardly likely to lift the spirits of the rebellious militia.

The weather now became severe, with heavy rain and a bitter wind. The canvas tents – each housing ten men – began to leak like sieves. The rain was then followed by heavy frosts, chilling the men to their bones and turning the soggy campsite into a skating rink. Many of the men lacked coats, which had been promised but never supplied by the quartermaster. But it was getting harder to desert so far from Cincinnati, which at least solved one of St. Clair's problems. But the cold and damp were giving him problems of his own and he was suffering agonies from rheumatism.

The route through the dense, gloomy forests filled the men with foreboding. They felt that there was an enemy behind every tree and that they were constantly being watched. Overhead the rain fell, occasionally turning to hail and snow. Stragglers began to be picked off by the Indians. At night St. Clair took extra precautions against Indian attack, ordering every man to stand to his post just before dawn each morning – the favoured time for Indian attacks. When violent storms hit St. Clair's camp early in November it was the final straw for many of his militiamen, who had been grumbling almost since they left Cincinnati. Seventy of them decided to desert, and threatened St. Clair that if he tried to stop them they would attack his supply convoy moving up from Cincinnati. In despair, St. Clair had to let them go. He also sent some of his best regular soldiers back to defend the convoy in case they decided to attack it.

Snow had begun to fall as St. Clair pressed forward into the heart of hostile Indian country, along the Wabash River. His scouts were travelling less distance away from the camp each day and now they failed him completely. As a result of their failure to reconnoitre properly, the Americans made camp just two miles away from a large encampment of Indian braves. St. Clair, unaware of their proximity, considered surrounding his camp with trenches or breastworks, but his men were too cold and exhausted to bother. Even when three senior officers received reports that there were large groups of Indians in the forest they decided not to worry St. Clair with this alarming information.

On previous days the troops had been called to their posts ten minutes before daylight and then stood guard until sunrise before settling down for breakfast around the campfires. But on this one day – as fate would have it – the system was altered. Reveille was blown rather earlier than usual and in compensation the men were released earlier, retiring to their campfires half an hour before sunrise. With the men disarmed and seated around their cooking kettles they were helpless when the Indians suddenly fell upon them with bloodcurdling shrieks and war cries.

The Kentucky militia took the full force of the Indian charge, fired a few shots and then fled like 'a shameless lot of cowards' back across the Wabash River, spreading confusion throughout the main camp. Finding cover behind trees and fallen logs, the Indians picked off the officers with remarkable accuracy. The American artillery fired round after round harmlessly into the trees while the Indians shot the gunners down one by one. Few of the new recruits even knew how to use their muskets properly, misloading, misfiring, and shooting friend and foe indiscriminately in their panic. According to one officer present, the enlisted men did much slaughter on the 'twigs and leaves of distant trees', but very little on the Indians.

Even though the Americans outnumbered the Indians by 1,400 to 1,000 men there was little orderly resistance. St. Clair was quite incapable – through ill-health and fatigue – of commanding his troops. In the words of Colonel William Darke, 'A general, enwrapped ten-fold in flannel robes, unable to walk, placed on his car, bolstered on all sides with pillows and medicines, and thus moving to attack the most active enemy in the world was . . . tragi-comical . . . indeed.' Suffering so badly from gout that he was unable to walk, St. Clair was twice lifted onto a horse, only for the animal to be shot down by Indian fire. As the Indians moved towards him with tomahawks and scalping knives at the ready the general underwent a transformation – his gout and rheumatism were forgotten. As he himself said, 'I could wait no longer; my pains were forgotten, and for a considerable time I could walk with a degree of ease and alertness that surprised everybody.' It was a miracle cure; its name was pure, naked terror.

While St. Clair was undergoing his metamorphosis, Colonel Darke heroically charged the Indians and drove them back across the Wabash, gaining time for others to escape. St. Clair continued to lead a charmed life. Eight bullets passed through his clothing without harming him. Meanwhile, the Indians were continuing to massacre their foes. One American officer was scalped as he sat against a log wounded and was later seen with 'his head smoking like a chimney'. The Indians killed General Butler and ate his heart because of their respect for his great courage. The women suffered a dreadful massacre, but not before they had shown more courage than the cowardly militiamen. In fact, the women drove 'the skulking militia and fugitives of other Corps from under wagons and hiding places' and back into the fight. But only three women escaped with their lives.

Desperate charges by a revitalized St. Clair, helped by the heroic Colonel Darke, broke the Indian ring around the camp to the south, allowing some 200 soldiers to escape from the trap. But once they had gone there followed a dreadful massacre, with the frenzied Indians torturing their victims in an almost infinite variety of ways. Dancing and howling at the screams of prisoners roasting at the stake, they eviscerated some while flaying others alive and hacking their limbs away one by one. Babies and young children were swung round and smashed against tree trunks, while some of the women were ravaged with pointed wooden stakes or 'cut in two, their bubbies cut off and burning'.

The Indians declined to pursue the survivors, and St. Clair reached Cincinnati with news of the disaster. Thirty-five officers had died along with 622 soldiers and civilian employees. Total casualties of 918 included wounded soldiers as well as dead camp followers and women whose actual numbers had not been officially recorded. A committee of the House of Representatives was appointed to investigate the disaster and reported that St. Clair should be completely exonerated of any blame. The troops he had been given had been quite unsuited to their task, the quartermaster and contractors inefficient and the weather unexpectedly harsh. But this was little short of a whitewash. Though undeniably a brave man, St. Clair was quite unfitted through health and experience to lead such inexperienced troops into Indian country. His strict discipline only irritated the militiamen and drove them to mutiny and desert. His procrastination caused delay after delay, making it certain that the expedition could not be completed before winter. Through personal faults he failed to weld his officers into an efficient team, and as a result allowed his army to degenerate into a demoralized rabble. These criticisms may seem severe when one considers St. Clair's age and physical condition. But one must be severe in assessing the reasons for a disaster of this magnitude. He should have made it clear to Knox and Washington that he was unfitted for the command, and supported the appointment of a younger man. His poor health lost him the respect of his men and contributed to the collapse of discipline on the march, while his campaign of floggings and executions simply alienated the troops yet further.

THE BRITISH EXPEDITION TO RIO DE LA PLATA (1806–7)

Dreams of 'El Dorado' – South America's land of gold – have exerted a powerful pull on governments and individuals since Columbus first reached the New World. Yet never in three hundred years did this dream lead a government to lose its head so completely as was the case in 1806 when Lord Grenville and his colleagues fell for the 'three-card trick' served them by one of their own naval officers, Commodore Home Popham. Popham, more entrepreneur than sailor, had devised a plan for an expedition to South America to divert its wealth into British hands. With the backing of London merchants (previously barred from direct trade with South America), and with just 1,700 soldiers under Colonel Beresford, 'borrowed' without permission from the Cape garrison, Popham set off like a latter-day conquistador, to break the Spanish hold on the province of Rio de la Plata. Bounded to the west by the towering Andes mountains and to the north by Brazil, Rio de la Plata was a vast area including almost all modern Argentina, as well as the republics of Uruguay, Bolivia and Paraguay. The city of Buenos Aires itself was the largest in South America, and its population of over 70,000 people were almost certain to regard the British troops as foreign invaders. Yet none of this dented Popham's confidence, buoyed as it was by intelligence reports that others might have found less reassuring. In one of these an American sea captain, well-wined and dined, had told Popham that the people of Buenos Aires were so hostile to Spain that they would welcome the British as liberators. In another, an English ship's carpenter had reported that the city of Montevideo – guarding the mouth of the Rio Plata – was weakly garrisoned, with crumbling defences. On this slim evidence Popham was prepared to invade an area as big as Western Europe with fewer troops than might have been needed to patrol the public houses of Portsmouth on a Saturday night.

Impossible as the mission seemed, luck initially rode with Popham, and when he and Beresford arrived off Buenos Aires on 24 June 1806 they took everyone by surprise, forcing the city to surrender without bloodshed. The Spanish Viceroy, the Marques de Sobremonte, was at the theatre celebrating his daughter's engagement when Popham was sighted, and he immediately panicked and fled inland after seizing much of the gold from the state treasury. The people of Buenos Aires were just as surprised: one Spanish lady only learned of the invasion when she found a British officer asleep on her sofa – with his dirty boots on one of her priceless quilts.

It seemed that Popham had been right after all, and to prove it he sent back to Britain over a million dollars in prize money that Sobremonte had not been able to carry with him. But appearances were deceptive. The truth was that the British had just 1,700 infantrymen, eight cannon and only sixteen cavalry to control the whole of Rio de la Plata, and as soon as the locals realized how weak their conquerors really were, swarms of gaucho light cavalry and Creole irregulars began to close in on the city. Crisis point was reached when the local commander, Captain Santiago Liniers, attacked Buenos Aires with his force of regular troops. Urged on by their Catholic priests, the entire population of the city took up arms against the British. Beresford tried to get his men back to the ships, but before he could do so the people – fighting from the flat roofs of their houses – fell on his meagre force and forced them to surrender. While Beresford and his men were marched ignominiously inland as prisoners, Popham, from his ship anchored in the river, could only fume as he saw his dreams of a new 'El Dorado' fading before his eyes.

However, back in Britain, gold fever was sweeping the country. Merchants and traders of all kinds took Popham's claims of a 'New Arcadia' at face value and began shipping cargoes off to Buenos Aires in the hope of boundless riches. On 20 September, the booty from Popham's conquest was paraded through the streets of London in grand procession, bedecked with ribbons, flags and – prominent above all – the word 'Treasure', spelled out in gold letters. With the public in such a mood the government had no option but to back Popham's piracy and send him 4,000 reinforcements under Sir Samuel Auchmuty. Excited by the ease with which Popham had plucked such a rich plum from Spain's dying grasp, the Minister for War, William Windham, came up with one of the most astonishing – and idiotic – plans ever to emanate from the brain of a British Minister of War. Robert Craufurd, commonly known as 'Black Bob', was to be sent with a further 4,000 men to effect a landing on the west coast of South America, subdue the province of Chile, contact Beresford at Buenos Aires and then build a chain of military posts between Valparaiso in Chile and Buenos Aires. Geography can hardly have been Windham's strongpoint or else he would have realized that the distance between the two provinces was more than nine hundred miles and the terrain nothing less than the Andes mountain range. One might have pardoned this as a temporary aberration had it not been immediately followed up by a plan from the Prime Minister Lord Grenville himself for an attack on Mexico from both the east and the west coasts simultaneously.

While His Majesty's ministers were apparently taking leave of their senses, Auchmuty had arrived at the Rio Plata, only to discover that, far from being in control of Buenos Aires, Beresford and his entire force were prisoners somewhere in the interior. Too weak to recapture the capital, Auchmuty decided to make a base for himself by taking the city of Montevideo, which according to Popham's reports was weakly defended. As usual Popham's optimism was baseless. The city's defences turned out to be in an excellent state of repair and the defenders had over a hundred pieces of artillery. Though heavily outnumbered in both men and guns, Auchmuty's disciplined British troops forced a breach in the walls and took the city at the point of the bayonet. The enemy losses amounted to over 1,300 killed and wounded with 2,000 taken prisoner. While Auchmuty was busy mopping up pockets of resistance he was surprised by the appearance of Beresford and Colonel Pack, who had managed to escape their captors. Pack agreed to stay and help Auchmuty but Beresford had had enough of South America and sailed off to greater things on the battlefields of Spain.

In England the government had decided to send Lieutenant-General John Whitelocke, a foul-mouthed martinet with a pathological fear of getting wet, to take command of the forces at Rio de la Plata, and to act as governor after the province had been subdued. Despised by the rank and file for his coarse manners, and constantly at odds with his second-in-command, Major-General Leveson-Gower, Whitelocke was a most unwise choice for such a delicate operation. By the time he arrived at Montevideo it was winter and quite the wrong time

The abortive British attack on Buenos Aires under General Whitelock, 5 July 1807. Trapped in narrow streets and attacked from the flat roofs of the surrounding houses, many British troops were forced to surrender.

for campaigning. But Whitelocke, lacking local knowledge, did not grasp this, and he set about addressing the problem of how to approach Buenos Aires by river. Beresford had managed it by landing at Point de Quilmes, only eight miles from the city, but Whitelocke decided that the troopships would not be able to carry the troops further than a point 29 miles away from Buenos Aires, as the water was too shallow. This bizarre decision condemned Whitelocke's force to a dismal tramp through hostile country, against irregular forces who knew the land well, and over terrain that presented considerable and unexpected difficulties. Between his point of disembarkation at Ensenada and Buenos Aires, the area beside the river was low-lying and marshy to a depth of four miles, and no more than two feet above water level. To add to this inconvenience, the damp and cold played havoc with the men's morale. The lack of trees made it impossible for them to find fuel for fires or for shelters, and they were thus rarely able to eat hot food or sleep in dry blankets. To make matters worse, Whitelocke was obsessed with the harmful effects of rainfall and inflicted on everyone a savage regime ofmarching to find cover the moment a cloud appeared in the sky.

Whitelocke formed up his army into four brigades – the van commanded by Craufurd, and the others by Lumley, Auchmuty and Colonel Mahon – while a garrison of 1,300 men was left in Montevideo. Lord Muskerry, who knew the country between Ensenada and Buenos Aires better than anyone in the British force, was deliberately chosen to command the garrison at Montevideo after expressing the view that anyone attempting a landing at Ensenada in midwinter must be mad. Piqued by this implied criticism, Whitelocke also left behind two companies of the 38th – Muskerry's regiment – and the 47th, the best soldiers in his command.

On 28 June, after a peaceful journey up the Rio Plata, the disembarkation at Ensenada got underway and the soldiers' suffering began. In their attempt to occupy the heights that ran alongside the river at a distance of four miles, the British troops were forced to cross a swamp. As more and more of them tramped through it became a filthy treacle that sucked men, horses and baggage down at every step. The guns stuck fast and could not be dragged clear even by double teams of horses, while the provisions were ruined by sea water. The local horses, still unbroken, would not bear a saddle and broke away, kicking madly and tipping their packs into the ooze. Of eight tons of biscuit disembarked just one ton reached the army intact, while mule-carts were equally unable to transport the supply of spirits. It needed the labour of hundreds of sweating redcoats to rescue the guns.

After two days the horrors of the swamp were past and Whitelocke was able to order the advance towards the village of Reduction, with Craufurd's division in the lead. Faulty staff work meant that many troops did not receive orders to bring three days' rations with them, while others had lost theirs in the march through the swamp. In consequence the troops had to go without food for at least two days. So obsessed was Whitelocke with getting his men under cover before the winter rains broke that on one occasion he ordered his entire force to march on, leaving behind their half-cooked dinners of captured oxen on the ground uneaten, all because he thought he felt rain.

Yet some of Popham's luck seemed to have rubbed off on Whitelocke, for as he marched towards Buenos Aires he was suddenly presented with the chance of an almost bloodless victory. Before reaching the city the British knew they must cross the swirling waters of the Chuelo River, whose only bridge was guarded by 9,000 Spaniards with 50 guns. By luck Craufurd's scouts observed a retreating enemy cavalry squadron crossing the river by a secret ford and Craufurd ordered his own advanced party to follow them. The last obstacle to the British troops was removed at a stroke: Craufurd had Buenos Aires at his mercy. Acting on his own initiative, he attacked the enemy troops in front of him and ordered a general pursuit into the unguarded outskirts of the city. On reaching a large open space known as the Corral – generally used as the city slaughter yard – Craufurd was fired upon by a Spanish cannon, whereupon his men charged the nearby hedges and cleared them of enemy infantry. At this moment Whitelocke intervened and spoiled everything, ordering 'Black Bob' to withdraw at the moment of victory.

Whitelocke, meanwhile, was moving up with the main body of British troops, prior to besieging the city. He had asked Gower to prepare a plan of attack, but what Gower finally came up with was, to say the least, eccentric. He proposed dividing the British force into thirteen separate parts and entering Buenos Aires down thirteen different streets. No arrangements were made to allow communications between the separate columns; each was to be entirely on its own. One column was instructed to seize the commanding buildings in the Plaza de Toros, while the others were to push on to the river's edge, capturing the buildings there and forming up on the roofs. What was supposed to happen next was uncertain, since Gower would not be with any of the columns and Whitelocke would be outside the city in the role of a passive spectator.

Why Whitelocke accepted such a plan is difficult to understand. On his arrival at Rio de la Plata he had pointed out to Craufurd that the construction of the local houses – with flat roofs surrounded by parapets – made them ideally designed for defence against just such an attack as Gower envisaged. He had added that he would never expose his own troops to the unequal contest of street fighting against irregulars on rooftops. His own plan, as far as one can tell, would have involved a heavy and prolonged bombardment, possibly in conjunction with the naval forces on the river, before his assault troops were committed. Why then did he adopt Gower's hare-brained scheme?

Whitelocke, it seems, was becoming uneasy about the condition of his men, who were tired, short of provisions and exposed to the rigours of the rainy season. He had found it very difficult to maintain links with the fleet and to transport supplies to the troops on the march. Rather than find a solution to the supply problem, he preferred to risk everything on winning a quick victory. Once the city was captured his problems would disappear, as his troops would be able to shelter from the rain and feed in some comfort.

Gower instructed his officers that the attack would begin at noon, but Auchmuty pointed out that his officers had had no time to acquaint themselves with the plan, nor to reconnoitre the ground, and that midday was an extraordinary time to assault a heavily populated city such as Buenos Aires. Reluctantly Gower was forced to agree and the attack was postponed until first light the next morning.

Everywhere officers tried to make some sense of Gower's plan. The fact was that the enemy were bound to know the layout of their own city better than British troops entering it for the first time, and ambushes were certain. Gower should have realized that besides the fire from regular soldiers, the British would be attacked from the rooftops by missiles of every kind, from grenades to chamber pots, tables and chairs to porcelain vases, thrown by almost every able-bodied civilian, man, woman and child. Without some form of artillery bombardment the British troops were throwing away their technological advantage and fighting an enemy who had the advantage of the high ground. Even more incredible was the decision taken by Gower to order the troops to remove the flints from their muskets before setting off, so that the city could be taken at bayonet point. Yet if the momentum of their attack broke down and they were forced to take cover the British soldiers would be helpless. While the British made their own task virtually impossible, the Spanish defenders did what they could to help. Cannon were placed at the ends of the streets and trenches dug across the main avenues leading to the Great Square. Houses were barricaded and missiles were assembled on the flat roofs to hurl on the advancing British infantry. Priests had roused the people to a religious fury and everyone was willing to play a part in defending their city. There may have been as many as 15,000 Spanish and irregular troops in Buenos Aires commanded by Liniers, as well as thousands more ordinary citizens who would join in when the time came. The British assault was to consist of 5,000 men, with something over 1,000 men kept in reserve. Each of the thirteen columns was to be led by two corporals equipped with crowbars, presumably to dismantle barricades.

Just after dawn the advance began. Despite its population of 70,000 the city was deathly silent, with scarcely even a scavenging dog to be seen. The sound of marching soldiers and cannon being dragged over the cobbles echoed in the stillness. On the left Auchmuty advanced a mile without meeting any resistance, until two cannon suddenly opened up on his columns, followed by a heavy fire from hidden musketeers. Simultaneously, firing burst out from many parts of the city as if by prearranged signal and there were shouts and screams of anger from the rooftops as Negro slaves, women and even children joined the attack. Within minutes Auchmuty's force was decimated. To the south of Auchmuty, Lumley's brigade, the 36th and 88th, had encountered strong opposition. The 36th fought its way through to the river bank and hoisted its colours in a tall building there. However, this action only brought down on them a hail of fire from cannon in the fort and marksmen in the surrounding houses. Lumley could do no more than hold his position.

The 88th were in even more trouble. Their left column under Major Vandeleur advanced under a hail of bullets, grenades, stones and household garbage, while they were raked by cannon in the Great Square. With great courage Vandeleur led his men over a sandbag defence in the road only to discover that they were trapped, with no outlet to the river. Fighting their way into a nearby house it soon became apparent that they had no hope of escape and so were forced to surrender. The Light Brigade had advanced in two columns, one of 600 men under Colonel Pack, the other under the personal direction of 'Black Bob' Craufurd, but both were forced to seek cover in the Convent of St. Domingo. Here Craufurd was besieged until noon when a Spanish officer, showing a flag of truce, brought a message from Liniers, which Craufurd assumed was a Spanish capitulation. Instead it was a demand for his surrender. Although Craufurd rejected it, by 3.30 pm it was apparent that his position was hopeless and he too was forced to ask for terms.

Throughout the fighting Whitelocke had simply paced up and down at his headquarters awaiting news. In fact, no reports reached him because none of the commanders in Buenos Aires knew where he was. So confused was the fighting in the city that most of the British troops were trapped in a maze of narrow streets where they were shot at from all sides, or holed up in churches or houses for protection. Once they entered the city Whitelocke's troops were beyond his control. At 9.00 Whitelocke sent his ADC, Captain Whittingham, towards one of the central streets, but he was driven back by heavy fire. Whittingham later climbed to the top of a nearby house and was able to report British colours flying in various places. Five more hours passed before Whitelocke called for a volunteer to try to gain news of Auchmuty. Whittingham, with an escort, was able to fight his way to Auchmuty and back with news of mixed fortunes. Nobody seemed to know what had happened to Craufurd until a letter arrived from General Liniers which dashed all Whitelocke's hopes. Liniers wrote that he had captured Craufurd and well over 1,000 men but was prepared to free not only them but also the survivors of Beresford's command if Whitelocke would remove all his troops from the province.

By the time Whitelocke and Gower met Auchmuty at Plaza de Toros to assess their situation, it was apparent that British casualties had been very heavy – nearly 3,000, with over 400 killed. On the other hand, the Spaniards had themselves sustained heavy losses and the British were holding several strategic positions on both sides of the city. It was only a matter of time before the city fell, but Whitelocke was not to be granted that time. His troops no longer had any confidence in him. Indeed, General Craufurd was so furious with his commander that he had ordered his own men to shoot him if they got the chance. Everywhere, troops began scrawling graffiti on walls suggesting Whitelocke was 'either a coward, or a traitor, or both'. Yet Whitelocke had no real choice but to agree to a withdrawal. Even if he had completed the capture of the city he could not hope to control the whole province without committing the British government to a wholly disproportionate military effort. He signed the agreement with Liniers to restore all prisoners and to evacuate the province within ten days. A proposal for liberty of commerce for British traders was rejected by the Spaniards. So ended Popham's dreams of commercial empire.

When the news reached London the reaction was predictably hostile. It was heightened by fears of financial loss. Whitelocke's greatest critics were those who had foolishly believed Popham's promises of unlimited wealth and had staked their life savings in this South American venture. Nothing less than a court-martial would satisfy such a wave of public indignation. Whitelocke was found guilty of mishandling the military operations by ordering his troops to attack Buenos Aires with unloaded arms and failing to control or support the columns once inside the city. The sense of national humiliation felt by the British public undoubtedly influenced the findings of the court, and little consideration was given to the difficulties Whitelocke had faced. After all, it was most unlikely that the Spanish colonies, having freed themselves from the corrupt rule of Spain, would have immediately accepted rule by Britain. Only a form of independence under British protection might have been acceptable, and it was precisely this sort of commitment that the British government had forbidden Whitelocke to make.

Whitelocke was made the scapegoat of an impossible mission foisted on him by a government that was acting foolishly. He alone suffered ruin: Popham emerged from the fiasco to become an admiral, while Beresford and Craufurd became two of Wellington's most able lieutenants during the Peninsular War.

It is promotion by purchase which brings into the service . . . men who have some connection with the interests and fortunes of the country.'

The Duke of Wellington, defending the system of purchasing commissions

The Duke of Wellington was to the first half of the 19th century what the Duke of Cambridge was to the second – the authoritative voice of outdated military thinking which condemned Britain to a century of military incompetence.

THE BATTLE OF ASPERN-ESSLING (1809)

Crossing a broad river in the face of the enemy is a dangerous military manoeuvre as the Earl of Surrey found at Stirling in 1297 (see p. 103) and General Ambrose Burnside at Fredericksburg in 1862 (see *The Guinness Book of Military Blunders*, pp. 116–17). Nor was the Danube any easier when Napoleon tried to cross it during his Austrian campaign of 1809. Napoleon, of course, was hardly a blunderer of Burnside's class, and he had successfully negotiated such problems many times in the past. Yet in crossing the Danube near the villages of Aspern and Essling in May 1809 he made such elementary mistakes that he suffered his first ever defeat in battle.

Having vanquished the Austrian commander, the Archduke Charles, at Eckmühl on 22 April, Napoleon decided that he must follow up his victory and eliminate the main Austrian army before it linked up with the Archduke John's troops in the south. Between him and Charles, however, was the river Danube, swollen by spring floods. Leaving large detachments under Bernadotte and Vandamme to guard against the Austrians recrossing the river, Napoleon concentrated his attention on securing a bridgehead as soon as possible. But the dangers were greater than Napoleon realized. So impatient was he to complete the destruction of Charles's army that he ignored warnings that unless careful preparation went into the bridging of the Danube the venture was likely to prove a disaster. In the first place, unless he was able to prevent it the enemy upstream would float down fireships and other obstacles to smash his bridges. In addition, the Danube was a river prone to sudden spates of floodwater in the late spring which could easily overwhelm his flimsy pontoons. Yet Napoleon refused to listen to these warnings, so confident was he that a complete victory over the Austrians was within his grasp.

Napoleon's chief engineer – General Bertrand – carried out a careful reconnaissance of possible crossing points. He decided on a spot near the island of Lobau and began to build a bridge to the island. But because Napoleon was so obviously in a hurry, Bertrand cut corners and built the bridge without protective pallisades and without flotillas of manned river boats, which would have headed off anything the Austrians floated downstream to damage the pontoons. Assuming that the Austrians were far away to the north, Napoleon had become careless and his overconfidence persuaded his officers that they had nothing to fear from an enemy so obviously in flight.

The usually cautious Berthier was also at fault, drawing up a poor plan for the crossing to Lobau. Dismissing the threat from the Austrian army, and convinced that his main job would be to hunt down a broken foe, Berthier placed numerous light cavalry in the vanguard of the crossing. Such forces were excellent at pursuit, but hopeless to resist a powerful counterstroke. Napoleon and Berthier were underestimating their opponents and would suffer for it. The Austrian army had not broken up and fled northwards as they had expected. Instead it had withdrawn in good order and was entrenched nearby.

As the crossing went on Napoleon became increasingly uneasy about his pontoon bridge to the island of Lobau. If it broke, he knew that his army would be split on either side of the river. He was

At the battle of Aspern-Essling, 21 May 1809, the Austrian Archduke Charles became the first general to defeat Napoleon. But he failed to capitalize on his opportunity to destroy the French army, and suffered a decisive reverse at Wagram a few weeks later.

exposing himself to an enemy counterstrike and he knew it, yet surely the Austrians were on the run. He was convinced of it. And yet. . . .

Overnight the Danube rose three feet and 'water-borne missiles' – fireships, logs and floating mills – continually battered against the leaking pontoon bridges. Just as Napoleon had decided to recall his troops from the right bank, control of events was suddenly ripped from his hands as the Austrians attacked. Had the Austrian commander realized the hopeless position Napoleon was in he would have destroyed the French troops on the right bank and possibly brought the Corsican's career to a premature end. But he could not believe his luck and was hesitant, thereby missing his chance.

For the first four hours of fighting fewer than 23,000 French troops were facing over 100,000 Austrians, the latter vastly superior in guns and with the prospect of heavy reinforcements. Never before in his dazzling career had Napoleon faced such a catastrophe. And as if they knew it his magnificent troops fought as never before. Survival rested on the arrival of Davout's Corps across the pontoon from the left bank of the river. But once again the bridge was broken and Napoleon was forced to pull back his men from the villages of Aspern and Essling where they had fought so tenaciously. He was beaten and he knew it, but things could have been far worse. He had saved his army from potential disaster. Casualties on both sides were enormous – a combined total of 46,000 killed and wounded – yet Charles had missed his chance of driving the exhausted French into the river.

Napoleon's conduct of this battle had 'bordered on madness'. He had behaved as if he could control the elements of wind and tide as well as the forces of man. He had fought a battle without any knowledge of the Austrian whereabouts, without securing his passage of the river, and without assembling his whole strength on the island of Lobau. Yet in a sense the Archduke Charles' failure was even greater. To have the *Grand Armée* divided on both sides of a river in flood, and with a substantial part of it trapped on an island in the middle of the Danube, would have been the dream of a Wellington or a Blücher. Surely there was no escape from such a trap even for Napoleon. Yet Charles had none of the stuff of greatness about him and, more importantly, Napoleon would not underestimate him next time. Six weeks later the French emperor decisively beat the Austrians at Wagram, ensuring himself a further six years of supremacy in Europe.

'If your General keeps a girl, it is your duty to squire her to all public places and to make an humble third of a party at whist or quadrille; but be sure never to win.'

Advice to staff officers from the Duke of Cambridge, 1857

Officers needed to be trained in etiquette, notably in dealing with the minor peccadilloes of their commanders. This was far more important than the technical and scientific training available at staff colleges, regarded by the Duke of Cambridge as more harmful to the aspiring officer than houses of ill-repute. If the Duke had not existed one would have thought him an invention of *Punch*.

THE SIEGE OF BURGOS (1812)

Even as distinguished a military commander as the Duke of Wellington had his 'off days'. The Iron Duke's least successful operation was probably the siege of Burgos during the Peninsular War, in September and October 1812. Judged by his own high standards, his performance on this occasion was inept in every way and there were moments during the siege and in the retreat that followed when Wellington seemed to be in the grip of a mischievous and malignant fate.

He began badly by belittling the task before him, describing the ancient Castillian city of Burgos as nothing more than the kind of hill fort he had captured many times during his military career in India. In fact, Burgos, the depot of the French army in northern Spain, was a strong fortress built on the ruins of stout medieval fortifications. It was well supplied with food and all kinds of ordnance, and its garrison of 2,000 men was ably led by a master of siege warfare, General Dubreton. How Wellington expected to take the city with only five engineer officers and eight sappers, and a total siege train of eight cannon, is beyond comprehension. And when it is recalled that the British had over 100 siege guns in Madrid which were not sent for, and that Admiral Home Popham offered to send heavy cannon from the fleet, only to have his offer rejected, one is entitled to wonder whether this was the Duke of Wellington known to history and not some imposter.

To compound the folly of rejecting the siege guns Wellington tried to take the fortress by infantry assault. By his own admission he was not very confident about this tactic, but was still willing to squander his men in futile storming parties. When this failed – as the humblest private soldier could have told him it would – an atmosphere of gloom and defeatism descended on the British camp. When an officer carrying Wellington's plans in his pocket was shot and taken by the French, it was as if the Almighty himself had deserted the British cause.

When the siege of Burgos began, the target of the first British attack was an outlying structure known as the Hornwork. Attacking at night under a full moon and without any preliminary bombardment, the British and Portuguese infantry suffered 421 casualties before getting some men inside and forcing the French to abandon the position and flee into Burgos. Encouraged, Wellington ordered his men to attack the outer walls of the fortress itself. But no breach had been made in the walls and a further 200 casualties were sustained to no purpose. Next the engineers placed a thousand-pound mine under the foundations of the north-west wall of the fort. It was not until it exploded that they realized that they had made an embarrassing miscalculation: they had placed it under the walls of the old medieval castle, rather than the modern fortification. Apart from shifting tons of ancient masonry the mine achieved nothing. Meanwhile, Wellington's artillery – all eight guns – had begun a bombardment of the fortress. However, his 24-pounders proved inefficient, using so much powder that they had to be taken out of action. When an attempt was made to set fire to the main stores in Burgos with red-hot shot, it failed to catch light, though the shock of the explosions did cause a passing French commissary officer to lose his reason. Eventually the French defenders, tired by the squeaking of the British guns, roared back with their batteries and put the British gunners to flight, destroying most of their cannon in the process.

October came and Wellington could think of nothing better than to do more of the same. He ordered his engineers to plant another mine on the north-west walls, and succeeded this time in making a small breach. But the French were waiting for the storming party sent to exploit the breach and drove them back with a further 220 casualties. To compound Wellington's misery, Dubreton then launched a successful sally from the fort, occupying the British trenches, destroying some siege equipment, wrecking some of the British excavations and drinking the soldiers' grog.

And now the weather changed for the worse. Rain fell in torrents, flooding the British trenches and depressing everyone. Criticism of Wellington was widespread. He made one last assault – by infantry alone – which was contemptuously turned back by the French. Then he decided to raise the siege. When the engineers tried to destroy the Hornwork to deny it to the French, the mines failed to explode and they were forced to leave it intact. By 22 October, when he called off the siege, Wellington had lost well over 2,000 men against a cost to the French of just 300. At least the Duke's verdict was just: 'It was all my own fault.'

Wellington's obstinacy at Burgos had been symptomatic of a deeper malaise. One of his soldiers had noted, 'If ever a man ruined himself the Duke has done it. For the last two months he has acted like a madman.' Even his own doctor had been alarmed by the signs of Wellington's bad humour. War does strange things to people.

On the retreat from Burgos Wellington lost his grip on his army and the pressures of years of Peninsular campaigning began to tell. At the town of Torquemada, 12,000 British soldiers, unable to resist the temptation of the local wines, got themselves roaring drunk. The bodies of drunken men lined the roads as if a major battle had been fought. But the only enemy had been the newly fermented wines of Castile, which had proved too much for Wellington's battle-hardened veterans.

In the meantime things were going badly wrong with the army commissariat. Thanks to the incompetence of the temporary quartermaster-general, Colonel Willoughby Gordon, the army's food supplies had been sent via a road 20 miles to the north and completely missed the men for whom they were intended. This meant that for four days the soldiers got no food at all. In drenching rain and with a large French force pursuing them the British army became a disorderly rabble. Some got into squabbles searching for acorns, others raided the huts of local peasants or even shot at passing herds of pigs. One officer said he could have eaten his boots, another said he could have eaten Wellington's. In the ensuing chaos three thousand British stragglers were taken prisoner by the pursuing French. Sir Edward Paget, commander of the First Division, was captured by skirmishers, while three other generals disobeyed Wellington's order and followed a shorter route which the commander had specifically forbidden them to take. These disobedient generals led their troops into a dead end, finding their way blocked by a river in flood. Such insubordination could have resulted in the loss of two divisions to the French, and undone all of Wellington's good work in Spain. When asked what the 'Iron Duke' had said when he met the three errant generals, Lord Fitzroy Somerset (later Lord Raglan, of Crimean notoriety), replied, 'What did he say? Oh, my God, it was far too serious to say anything.' When Wellington met the officer in charge of the baggage he asked the wretched man what he was doing, and received the reply: 'I've lost my baggage.' Without stopping for a moment Wellington rode on, saying: 'I can't be surprised . . . for I cannot find my army.' He found it eventually – but, more important, he found himself.

THE ASSAULT ON THE REDAN (1855)

By May 1855 the Franco-British siege of the Crimean port of Sebastopol had been going on for some eight months, and its fall seemed just as far away as ever. The bombardments of the city attracted spectators to watch the regular pyrotechnic displays from the heights behind the Allied batteries. One officer described it as like a scene from 'Derby day', with wives, newspaper correspondents, gentlemen travellers, salesmen and off-duty officers picnicking and being entertained by regimental bands playing 'light airs' from the works of Schubert or Lanner. By general agreement, the best band came from the newly arrived Sardinian army, which played stirring selections from a number of operas. The many English ladies in the Crimea that summer, including the ravishing Lady George Paget – 'the belle of the camp' – were rather taken with the Italians in their bandit hats with large plumes of black cock feathers. Lady Paget's bold husband, Georgie, was near at hand as commander of the much-depleted British cavalry. Apparently the only use Lord Raglan could think of for Georgie and his horsemen was as 'special constables' to stop souvenir-hunters rushing into Sebastopol after it had been captured.

The French were just as keen on maintaining the spirit of gaiety. They arranged bathing parties, fishing trips for the more contemplative and race meetings for the more active. The Zouaves (Algerian recruits) organized their own theatre which put on outrageous farces. One observer, so overwhelmed by the number of foreigners watching the siege, opined that the whole world must be there, except perhaps the North American Indians. It was like a league of nations. But somewhere a war was going on.

If it was heaven on earth for the visitors, paddling in the sea or quietly dozing in the pleasant sunshine, it was hell for the inhabitants of Sebastopol, one of whose hospitals was later found by the British to have at least one corpse on every bed. Nor was it much fun for the French and British soldiers in the forward trenches, whose job it was to launch the attacks on important defensive points of the city such as the Malakoff tower and the Redan, a fortified position. So dreadful was the fighting around Sebastopol that the French issued their storming parties with a pint of wine each before they went into action. The British generals, as usual, were being as awkward as possible on the subject of dress. British troops were still wearing thick winter coats and trousers even though the weather was pleasant enough for the ladies to discard their shawls, while the ridiculous Sir George Brown was at odds with the Light Division because of his insistence that his men wear the stocks and high-buttoned collars that had tormented the British soldiers of the 18th century.

On 17 June the Allies began a massive bombardment of Sebastopol, prior to an assault on the Malakoff and the Redan. Unfortunately, the night before the attack the French commanders, Pélissier and Bosquet, had a furious argument which culminated in Bosquet being removed from the attack and sent to command the reserve troops. This caused chaos in the French ranks. Pélissier then proceeded to change the time of the attack by two hours, designating 3 am as the new start time. But he neglected to tell the British commander, Lord Raglan, of this change of plan. To make matters worse, the French assault was supposed to begin when a rocket was seen in the sky. By chance a shell, emitting sparks, passed across the sky 15 minutes before zero hour and was assumed to be the signal for action. In fact, neither of the two French columns was really ready, but they gamely attempted to make the best of things. If the French were not ready, however, the Russian gunners most certainly were; not a single Frenchman reached the ramparts of the Malakoff tower alive.

Raglan, watching the French being massacred, thought it only comradely for his men to be mas-

The British assault on the Redan, 17 June 1855, was not the heroic episode that this picture would suggest. Many British troops refused to obey orders and fled rather than face the Russian guns.

sacred too. He discussed the possibility of an English assault with Sir George Brown, who agreed that an attack should take place. This was coalition warfare – and he was showing solidarity with the French. During the British attack, which involved just 400 men and, it is almost superfluous to add, was a disaster, there occurred the following extraordinary sequence of events, involving Lieutenant Fisher of the Engineers. Fisher had been bringing storming tools in case a breach could be made in the Redan walls, but seeing that the attack was failing, he wanted to know who was going to bring back all the picks and shovels. He did not see why it should always be left to him. He met Colonel Yea and asked him, only to see him drop dead with a bullet through the heart. Next Fisher asked Captain Jesse, but before the latter could reply he was shot through the head. Fisher then approached two more officers who were both decapitated by cannon balls as he spoke to them. Another officer he approached ran away before he could ask the fatal question.

But one modest British success was achieved. Part of General Eyre's brigade – the 18th Regiment, an Irish one – had succeeded in breaking through a cemetery and occupying some outlying houses of Sebastopol. To their surprise they had found them inhabited by women and children. The Irishmen, first reassuring the Russians that they would not be harmed, settled down to enjoy themselves. With the battle raging all around them, they discovered a little piece of Erin in that God-forsaken place. Both officers and men from the regiment sat and drank the coffee that the Russian women seemed only too keen to make for them, and ate their pork rations, using fine books as plates. One house contained a piano and soon the lilting melodies of Irish songs were heard above the sounds of gunfire. The quaffing of liberal quantities of wine and spirits added even more life to the occasion and soon Irish soldiers were staggering around dressed in women's bonnets and dresses. There was also some looting, and in one bizarre incident two soldiers from the 18th had a stand-up fight over an article they both coveted, with other men from the regiment forming a ring around them. It was a curious microcosm of the greater struggle going on in the real world. The Irishmen did not hear of the failure of the assault until after nightfall. Nevertheless, the brigade returned largely unscathed, carrying their drunk and wounded comrades on doors they had ripped from houses to act as stretchers.

In spite of evidence to the contrary, the French blamed the British for the failure of the attack. The Italians, rather impudently, showed solidarity with the French by blaming the British as well. The truth was the French plan had been hare-brained in the extreme. Thousands of French troops had been ordered to run across open ground for a quarter of a mile, all the time under fire from Russian grapeshot and cannister shells, then climb a rampart of felled trees, all with their branches facing outwards, cross a ditch, cross a sunken road, and then clamber up a steep incline and climb a wall. It was scarcely surprising that the French suffered 3,553 casualties, 50 per cent of them dead. British casualties were far lighter, no more than 25 per cent, but the whole operation had been doomed from the start. As Garnet Wolseley – later one of Britain's most famous commanders – wrote, 'You can ask too much from even British soldiers.'

Lord Raglan was shaken by the failure of the assault. As the casualties were brought back he bent over one wounded officer on a stretcher, asking, 'My poor young gentleman, I hope you are not badly hurt.' The man, obviously in agony, spat at him and wished him in hell, blaming him for all the useless bloodshed. It was a savage comment and it struck the gentle soul of Raglan a blow from which he never recovered. Two days later, on meeting the British commander, an officer remarked, 'Good God! He is a dying man.' And indeed, within three weeks Lord Raglan was dead. As one particularly cruel epitaph ran, 'He was distinguished for nothing but his amiable qualities.' Not the epitaph of a warrior, perhaps, but not the worst comment on a man's life. His replacement, General Simpson, was sick, elderly and incompetent – in other words, he had all the qualities needed for the Crimean command . . . But other generals were leaving the Crimea in droves – or coffins. Sir George Brown went home – to general applause; some wags suggested he had gone home years before. With him went General Pennefather, while General Escourt died of cholera and General Codrington was taken seriously ill. They were not much missed.

As the summer ambled on – or dragged on – depending on whether you were spending your summer holiday on the heights above Sebastopol or shedding your blood down in the trenches, the fighting occasionally interrupted the social events. When the British were not attending races they

'What the devil are you reading those for? You are a horse artilleryman. What more do you want?'

A colonel of horse artillery

When Field Marshal Ironside was a young horse artillery officer he was once visited in hospital by his colonel. The colonel saw two military books on his side table and made the above remark. Later research showed that 95 per cent of British officers had never read a military book of any kind.

were playing or watching cricket. The day after the battle of the Chernaya the Guards Division fielded a team against the 'Leg of Mutton Club', a team made up of officers from various regiments. Between innings, and when the irritating sounds of heavy guns firing did not intrude, the players and their ladies enjoyed splendid food in good company.

While gentility held court on the heights, Generals Pélissier and Simpson were discussing a new assault on Sebastopol, planned for 8 September. Or rather, Pélissier talked about it while Simpson was apparently asleep, nodding forward and occasionally snoring loudly. The French took the nodding of his head as approval of the most outrageous of their demands. When the meeting was over the British learned that they were earmarked to attack the Redan again. The French had instructed Simpson to begin his attack as soon as the tricolour was flying above the Malakoff.

On 8 September zero hour had been set at midday, to baffle the enemy and to baffle military historians ever since. Within ten seconds of the French beginning their attack the tricolour was flying above the Malakoff. The British were caught entirely by surprise. Simpson was suspicious. How could the French have captured it so quickly? He waited 15 minutes. Then he sent an officer to the Malakoff to ask if the French really meant the British to attack. General MacMahon, sidestepping bayonet thrusts and ducking cannon balls, assured the British officer that he would be obliged if the British would do so. The officer returned and reported back to Simpson, who pondered on the problem. Suddenly a staff officer reported that the British had already decided to attack without orders. Apparently, as one senior officer reported later, the men could not be restrained and had rushed towards the Redan even before they could be equipped with ladders. Again the Russian gunners were waiting for them and again scythed them down in hundreds. Yet the survivors managed to take the outer walls of the Redan at bayonet point. The first waves of British and French troops had gained their targets and were hanging on grimly. The second wave of British infantry was now ordered to attack, but at once everything started to go wrong. Through gross mismanagement the second wave consisted entirely of young recruits, new to the Crimea. These men – many little older than boys – lacked the spirit of the more experienced soldiers in the first wave and refused to go out into the open under fire. Six hundred men crowded together, refusing to face the Russian fire, regardless of what their officers said. Some senior commanders were seen with swords bent out of shape from beating the backs of the laggards. So crowded was the mass of British troops that the Russians began tipping heavy stones onto them rather than firing. The attack fizzled out in an appalling rout. Old soldiers were apparently shocked at the cowardice they witnessed. Henry Clifford spoke of his heart breaking at seeing British soldiers, of whom he had been so proud, simply running away. And run they did, shouting 'The Russians are coming!' In the chaos hundreds more were shot down by the triumphant Russians. Casualties in just two hours were higher than in the dreadful fighting at Inkermann: 385 killed, 1,886 wounded and 176 missing. Henry Clifford questioned the logic of sending two divisions which had only just arrived in the Crimea to attack the Redan, and without any artillery support. According to Wolseley 3,000 men had been funnelled into an area of less than 2,500 square yards and then subjected to heavy fire from above. Simpson had blundered, but for once the failings of a commander had not been redeemed by the performance of indomitable British soldiers. A naval officer who witnessed the assault wrote:

> I am sorry to say the men behaved in a most cowardly and rascally manner, left their officers to be killed and failed entirely . . . The men held back or ran, the officers were all obliged to go to the front and were shot like dogs, not being able to get the men to advance.

The French had taken the Malakoff even if the British had run from the Redan – and it was enough for the Russians. To the astonishment of both commanders the Russians began evacuating Sebastopol. Pélissier joined Simpson and the coalition's success was sealed by a kiss on both the old gentleman's cheeks. Soon British soldiers were walking about inside the Redan. There they found the body of the only British soldier to reach the Russian breastworks – Ensign James Swift of the 90th Foot.

THE BATTLE OF CHINHAT (1857)

The Indian Mutiny saw many acts of heroism but far fewer examples of military skill. Too often the responsibilities of military command fell to those least equipped to assume them. Blunders were common and disasters frequent. Yet not every military débâcle was inevitable. There were occasions where rank stupidity contributed to the result, as at the battle of Chinhat, outside Lucknow, on 29 June 1857. Here a dispute between two men, Sir Henry Lawrence, Chief Commissioner for Oudh Province, and the Financial Commissioner, Martin Gubbins, brought about a battle that could never have been won and should never have been fought.

In 1857 Brigadier-General Sir Henry Lawrence was just 50 years of age but looked far older. Bowed down by work and responsibility, he was not an inspiring leader. During the early days of the siege of Lucknow there were some who agreed with Gubbins that only strong action would suppress the mutineers. Gubbins was particularly critical of Lawrence, writing that 'Sir Henry is no longer firm nor his mental vision clear'. Lawrence, in fact, was not a well man – his doctor had said that without rest his life would be in danger – and this allowed Gubbins to take liberties that would have been unthinkable if the general had been fit. He badgered Lawrence with demands that he take what troops he had in Lucknow and trounce the mutineers in the countryside around the city. On 29 June, Gubbins told Lawrence that spies had brought reports that 500 mutineers with one 'wretched gun' were at Chinhat, just eight miles from the city. When Gubbins received a second report that the figure of rebels was now 6,000, with many guns, he realized that this would spoil everything. Lawrence would never risk battle against such a force, which would outnumber him by ten to one. But Gubbins believed a show of force would be enough to scatter the rebels. He therefore concealed the second report from Lawrence, allowing him to take his men out to certain destruction. Sir James Outram later wrote that Gubbins' 'disregard of strict veracity' was to cost 300 lives and inflict on Lawrence an unnecessary disaster. Gubbins continued to goad Sir Henry, saying, 'Well, Sir Henry, we shall be branded at the bar of history as cowards.' Lawrence would have done better to have arrested Gubbins rather than rise to the bait. But rise he did and, gathering a force of no more than 600 men, half of whom – men from the 32nd Foot – had had no breakfast after a night of heavy drinking, he marched out to face the enemy.

Although he had planned to leave in the relative coolness of the early hours, delays meant that the men began their march in blinding mid-morning heat. The British soldiers, uncomfortable in their tight uniforms, soon wilted. Short of food and water, they were in no condition to face even 500 sepoys, let alone the 6,000 who awaited them in the vicinity of Chinhat. At first Lawrence met no opposition. Having reconnoitred, he considered falling back to Lucknow. Certainly the wretched condition of the British redcoats should have convinced him to retreat while he still could. Instead he asked the commander of the 32nd Regiment if his men were fit for action. They were clearly unfit and for a general to be asking a colonel to make his mind up for him was hardly impressive leadership. The commander of the 32nd – Colonel Case – knew only too well that his men were not fit for battle, but before he could reply the decision was taken out of his hands by the Regiment's previous commander, Colonel Inglis, who as Lawrence's second-in-command was unwilling to disappoint the general. Inglis replied that the 32nd were ready to advance, if ordered. Whether this was what Lawrence wanted to hear we will never know. But, having asked Case and Inglis to make the decision for him, he had no alternative now but to go on. And so the march continued until suddenly, as they approached the village of Chinhat, the 32nd were raked by heavy fire from sepoys concealed in orchards and groves of trees. The rebel leader, Barkat Ahmed, had nearly 6,000 men at his disposal and he deployed them with great skill, outflanking Lawrence and trying to cut him off from Lucknow. When he saw what was happening Lawrence ordered a general retreat, which quickly dissolved into a rout. His native gunners deserted to the mutineers and his Sikh cavalry charged off the battlefield in a fine display of *sauve qui peut*, leaving the British infantry stranded.

Alive to his folly – and to his own physical incapacity – Lawrence handed over command to Colonel Inglis with the words, 'My God, My God! And I brought them to this.' During the retreat to Lucknow the British suffered heavy casualties, many of the 32nd finding that their muskets had grown foul with disuse and would not fire at all. With a last desperate display of bluff, Lawrence ordered his gunners to deploy their pieces – even though they had no powder or ammunition – and hold flaming brands to their barrels. Faced by an apparently inflexible line of British guns the pursuing sepoy cavalry reined back, allowing the survivors of the 32nd to reach Lucknow. Yet, however heroic the retreat, there was no disguising the magnitude of the disaster. In all, Lawrence had lost 293 men killed and a further 78 wounded, out of his original 600. His first words on returning to the besieged city were naturally bitter: 'Well, Mr Gubbins has had his way and I hope he has had enough of it.'

In despair Lawrence later wrote, 'I look on our position now as ten times as bad as it was yesterday; indeed, it is very critical . . . unless we are relieved quickly, say in ten or fifteen days, we shall hardly be able to maintain our position.' Four days later Lawrence was killed when a rebel shell hit the apartment where he was sleeping. But Sir Henry had been a broken reed, prey to despair. Command now passed into younger, fitter hands and Lucknow survived not just one siege but two, and not fifteen days but more than 140.

'The sportsman has had more to do in winning our battles for us than anyone else, and what would have become of India unless our officers, military and civilian, had been inured to field sports and ready at any moment to take the opportunity of going out to kill a man-eating tiger, a rogue elephant, or any other dangerous beast.'

A.B. Wylde, 1885, during the Sawakin Campaign

The British way of war – short of a fourth for bridge or a slow left-armer to make up the XI, British officers had to make do with war to fill their social calendars.

TIMBUKTU (1894)

The charge of the sheep at Goundam in 1894 has failed to win a place in military history alongside that of the Royal Scots Greys at Waterloo or the Light Brigade at Balaclava, yet it was decisive for all that. The beasts, intermingled with Tuareg tribesmen, and running as if Lassie herself was behind them, completely overthrew the defences thrown up by the French troopers commanded by Lieutenant-Colonel Eugène Bonnier and wiped out his entire force. It was a disaster for France, but one that should never have happened, for Bonnier should not have been at Goundam in the first place.

Eugène Bonnier was typical of the ambitious, headstrong and insubordinate men who led France's army in the Sudan in the 1890s. His ambition was to make a name for himself by capturing the town of Timbuktu, and if that meant disobeying orders from his superiors then he was prepared to do it. But

Troops under the British general Sir Henry Havelock in action at Cawnpore during the Indian Mutiny, July 1857. Havelock fought his way through to Lucknow, relieving the Residency on 26 September.

the French government was firmly committed to a policy of retrenchment and wanted no more African territory. They even sent a new governor-general, Albert Grodet, to the Sudan, with express orders to stop Bonnier making any more territorial gains. Forewarned, Bonnier knew that he had to act fast and so he left the French headquarters just a day before Grodet arrived. With a force of 204 Senegalese riflemen, nine European NCOs and thirteen officers, plus two 80-millimetre cannon, he travelled up the River Niger in a flotilla of dugout canoes. Ahead of him, he sent Lieutenant Boiteux in two small gunboats, while a separate column under Major Joseph Joffre – later to be a portly commander-in-chief of France's armies in 1914 – set off overland to Timbuktu, carrying supplies.

When Grodet arrived at the French headquarters he was furious to find that Bonnier had escaped his clutches, and so he sent a message upstream ordering him to return. But the cunning Bonnier replied that he was merely carrying out a tour of inspection and would soon be back. In fact, Bonnier was having troubles of his own: Lieutenant Boiteux was no better at obeying orders than he was. Hoping to achieve the coup of taking Timbuktu himself, Boiteux had not waited for Bonnier's canoes to catch up but had steamed on towards Kabara, the port of Timbuktu, where he arrived on 28 December 1893. With just four Europeans and a few black sailors Boiteux wandered into the sleepy streets of Timbuktu and annexed it in the name of France. It had been a bloodless campaign, or so he thought, for while he was away from his steamers the local population attacked them and killed seventeen of his men.

Back at headquarters, Grodet was beside himself with rage. His enquiries had shown that Bonnier was deceiving him and had been planning the expedition to Timbuktu for months. He sent more messages, this time relieving both Bonnier and Joffre of their commands and demanding that they return at once. But before his messages arrived the local Tuareg tribesmen had decisively intervened in what had so far been a purely French problem.

On 10 January 1894 Bonnier's flotilla of canoes reached Timbuktu and he immediately arrested Boiteux, who was confined to quarters for 30 days – the luckiest imprisonment of all time, as it turned out. Leaving a small garrison of 50 Senegalese at Timbuktu – presumably to guard the irrepressible

A Tuareg tribesman c. 1900. An unlikely combination of angry Tuareg and frightened sheep overran the French encampment at Goundam in 1894, nearly nipping in the bud the career of a future marshal of France – Major Joseph Joffre.

Boiteux – Bonnier now set off overland to rendezvous with Joffre. He reached the village of Goundam, having captured about 500 sheep and camels from a Tuareg camp on the way. Instead of taking just enough sheep to feed his men, Bonnier insisted on sweeping the whole flock along with his column, slowing it down and stirring up a hornet's nest of resentment among the desert tribesmen. He spent much of 14 January skirmishing with bands of Tuareg on their camels and that evening he camped in a small clearing, surrounding the camp with a stockade of prickly bushes. He then ordered his men to stack their rifles in one corner of the enclosure, leaving the sheep outside. As his men slept the

Tuareg were able to creep up to the edge of the bushes before charging into the flock of sheep and driving the frightened animals into the French stockade, which was scattered to the four winds. The French soldiers, who unwisely were not sleeping with their rifles, were quickly overwhelmed. Few of them managed to reach the stacks of rifles in time and just one man – Captain Nigotte, who was knocked from his horse and managed to conceal himself in the bush – survived. When later questioned by Joffre, Nigotte claimed that the Tuareg had run alongside the sheep and had been almost impossible to detect until the last moment.

Bonnier's command was totally wiped out – the dead included thirteen European officers and NCOs and 68 riflemen. Bonnier had paid for a series of elementary errors with his life. His contempt for the Tuareg had made him underestimate the threat they posed, particularly after he had stolen their animals. His stockade was poorly constructed and surrounding it with so many animals had made it easy for the enemy to approach unseen. Stacking the rifles well away from the sleeping men – and letting them sleep even beyond dawn – was asking for trouble, and in France the press took the view that he had been incompetent and had got what he deserved. But, true as this might be, the greater problem was the insubordination of both Bonnier and Boiteux, which created a crisis where none had existed. French pride in their colonial army slumped when it was revealed how foolish and ignorant so many of the officers were. Joffre was sent home in disgrace, but he survived the shame well enough to serve his country again on a more important battlefield – the Marne, in 1914.

'We never worked as they work now. We hunted six days a week. I once hunted seven.'

Brigadier Monkhouse describing his training at the Royal Academy at Woolwich

Army training of the kind that would have been close to the heart of the Duke of Cambridge.

THE BATTLE OF OMDURMAN (1898)

The death of General Gordon in Khartoum in 1885 was an enormous blow to national prestige and caused the fall of Gladstone's government in that year. Herbert Kitchener, who had known Gordon well, took it very badly. Like almost everyone in Britain he ached for the day when the Sudan could be reconquered and Gordon avenged. It was appropriate therefore that when the decision was reached by Lord Salisbury's government to send an army down the Nile the command should be given to Kitchener. It was a decision that was to make Kitchener's career, winning him a peerage and an international reputation. Yet by creating a legend of Kitchener's infallibility as a warrior and a military thinker, the Sudan campaign served Britain badly. For Kitchener was an inept field commander whose deficiencies were frequently covered up by the achievements of others. Yet Kitchener had luck – the quality valued by Napoleon above all others.

It took Kitchener two and a half years to conquer the Sudan. It could have been done more quickly, but he had no intention of risking failure. His approach had all the inevitability of two thousand years of technical progress pitted against the shibboleths of chaos and ignorance. Kitchener was a child of his times, more interested in the nuts and bolts of the machines that helped him master the heart of the dark continent than in the achievements of Hannibal or Caesar or Alexander. He was an engineer trained at Woolwich, not a romantic like the young Winston Churchill who was to travel with the army. Kitchener's victory would be won before he even met the Dervish army and it would be a victory won by paddle-steamers, locomotives and maxim guns. There would be no room for heroics. Those who worked with Kitchener during his Sudanese campaign wrote of his obsession with technology. While Churchill had played with toy soldiers in his childhood, Kitchener had played – if he played at all – with trains. As a general he liked nothing better than driving a locomotive on the Egyptian tracks, or steering a paddle-steamer down the river. It is even reported that he liked to hammer rivets into the gunboats' hulls, though apparently his ADC had to mark each one he knocked in with chalk so that it could later be removed and hammered in properly.

By September 1896 Kitchener had laboriously cut his way through the desert to Dongola and here he proposed a halt of another six months in order to prepare for the next stage. In fact a year passed before he reacher Berber, after building a new railway through the Nubian desert. Nothing the desert could throw at him – not flash floods, cholera epidemics or sandstorms – could prevent him from creeping forward. At last, in March 1898, the Khalifa, who had succeeded to the temporal power of the Mahdi in the Sudan, decided to intervene. His son, Mahmud, led an army north to Atbara where he made it clear that he intended to fight the British. But this was poor strategy on the Khalifa's part – it divided his considerable strength and allowed Kitchener to chop it up bit by bit. Mahmud built himself a strong position at Atbara, with trenches and a *zariba* (a line of thorn bushes) to protect the camp. But Kitchener had 14,000 men, including four British battalions, and with these he took Mahmud's camp at bayonet point in just 15 minutes. He now decreed another three-month break, while he cabled London for another brigade of British regulars and other reinforcements. By August 1898 the new men had reached the Sudan and had been hastily transported by steamer to the vicinity of Omdurman. Kitchener now found himself in the comfortable position of having 25,000 men and 44 guns, as well as a flotilla of gunboats. Even 'King Chaos', as he was later to be called, would have to work hard to make a mess of this campaign.

On reaching Omdurman Kitchener made a mess of the Mahdi's tomb. A torrent of high explosive reduced the final resting place of Gordon's enemy to rubble. Kitchener was later to strew the bones of the holy man in the River Nile, keeping only the skull for an inkwell. Queen Victoria, on hearing of this desecration, told him to put the skull back. The rest of the bones were not recovered.

Just after dawn on 2 September 1898, Kitchener's cavalry patrols reported that the Khalifa's army – some 50,000 strong – was on the move towards where the British had established a fortified camp. At 6.50 am the British field guns, supported by the massed artillery of the gunboats, opened fire on the dense ranks of the Dervish army, inflicting huge casualties but not slowing the advance. The Khalifa then ordered two divisions of about 14,000 men, led by Osman Digna and Osman Azrak, to charge straight at the British camp. When they were still over a mile away the British regulars opened fire with their modern rifles; at half a mile they were joined by their Egyptian and Sudanese comrades with their older firearms. All the time the gunboats fired and the artillery tore gaps in the Dervish ranks. It was a massacre: not a single Dervish reached within 500 yards of the British camp. Even against such primitive tactics Kitchener still contrived to blunder, failing to get his British regulars entrenched so that they suffered quite unnecessary casualties from the random shots of the Dervishes. Ironically, the Egyptian and Sudanese troops, whose British officers had had the sense to dig trenches, suffered no casualties at all. Kitchener can hardly be praised for lining up his riflemen shoulder to shoulder as if he were an officer in the army of Frederick the Great. Having slaughtered several thousand tribesmen in a few minutes Kitchener's only comment was, 'What a dreadful waste of ammunition.'

Kitchener now reached the extraordinary conclusion that the battle was over. It had all been rather an anti-climax really. He could not see why everyone got so worked up about battles. Yet, unknown to him, the Khalifa had so far used only about a quarter of his army. Where were the rest? And why had Kitchener not taken the trouble to locate them? In fact, the Army of the Green Flag, some 20,000 strong, was somewhere to the north, while the Khalifa's own command, the army of the Black Flag, could have been taking tea in Cairo for all Kitchener knew. Convinced that he had won a complete victory, he ordered the 21st Lancers to pursue the beaten enemy. 'Annoy them as far as possible on their flank and head them off if possible from Omdurman,' ran the order to Colonel Martin, commander of the 21st. What happened next is as farcical as the Charge of the Light Brigade and

The charge of the 21st Lancers at the battle of Omdurman, September 1898. Winston Churchill rode with the Lancers and narrowly escaped death when they rode into a Dervish ambush.

almost as famous, thanks to the presence amongst the lancers of Winston Churchill.

Meanwhile, Kitchener had decided to break camp, line up his brigades and march off towards Omdurman, exposing his flank to an enemy force of at least 40,000 men of whose exact whereabouts he was still totally ignorant. This was rash, to say the least. The truth was that he had never commanded so many men before, even on exercises, and was frankly out of his depth. Whatever his excuse, it was one of the worst command decisions of the entire Victorian period, rivalling Lord Chelmsford's decision to divide his force at Isandhlwana in 1879 or Sir George Pomeroy Colley's to go to sleep on the top of Majuba Hill in 1881 (see *The Guinness Book of Military Blunders*, pp. 76–8).

While Kitchener was tempting fate, Colonel Martin was looking for death or glory. The 21st had long suffered the scorn of the army as being the only regiment without any battle honours whatsoever and Martin was determined to rectify this at once. He was unlikely to win a 'gong' for reconnoitring; nothing less than a full cavalry charge would do. Riding to the front and issuing the order to charge, Martin led his 400 officers and troopers in a mad charge across the desert sands towards a thin line of Dervishes. But as he breasted a rise he suddenly found to his horror that he had ridden into a force of 2,000 Dervish warriors commanded by Osman Digna that had lain concealed in a dry watercourse. As an example of how not to do things it was a masterpiece. Suddenly the exhilaration was replaced by a desperate battle to survive. In the hacking, slashing mêlée that followed, five officers, 65 men and 119 horses were killed or wounded. It was the biggest British loss of the whole battle and a perfectly pointless exercise in military anachronism. Why had cavalry been used when the infantry could have stood half a mile away and wiped the Dervishes out to a man without suffering any casualties? Furthermore, Colonel Martin gave little thought to exactly what Kitchener was going to do now that he had no cavalry to scout ahead. Martin rode back to report to Kitchener that he had carved himself a tiny niche in military history, as leading the last cavalry charge by the British Army. However, he had not done what he was asked to do and should have been court-martialled for rank stupidity. Instead, for his blunder he was made a Companion of the Bath, while three of his men got VCs.

Meanwhile, the two foremost British brigades formed up and marched out of the camp towards Omdurman. There was almost a holiday spirit now that the battle was over and the brigades vied with each other to be first to reach the enemy capital. Unfortunately, as the other brigades followed in line no one gave any thought to the rearmost brigade, an Egyptian one, commanded by Brigadier-General Hector MacDonald, a tough, God-fearing Highland officer, who was said to have knocked out a Boer at the battle of Majuba Hill with a straight right punch. Unknown to Kitchener, MacDonald was about to be attacked by fully 40,000 Dervishes, from the armies of the Green and the Black Flags. Fortunately for Kitchener, his reputation and the Empire, MacDonald was to display astonishing coolness under fire. Attacked from both sides he rode out in front of his men encouraging them and filling the air with all manner of Gaelic profanities. He sent Lieutenant Pritchard to tell Kitchener of his parlous position but the Sirdar took no notice whatsoever, ordering Pritchard to tell MacDonald to close up quickly and advance on Omdurman. Luckily one of the other brigadiers, Hunter, had a better grasp of the English language and managed to persuade Kitchener to take some positive action. Two brigades swung away from the march to Omdurman and came to MacDonald's rescue. While he was waiting to be 'saved' MacDonald did a little bit on his own account, smashing one massed attack from the west and another attack from the north, delivered simultaneously.

Scarcely bothering to look behind to see how MacDonald was getting on, Kitchener continued his march to glory. At 11.30 am he closed his field glasses and announced that the enemy had been given 'a good dusting'. This was something of an understatement. Nearly 11,000 Dervish bodies were found on the battlefield, and as many as 16,000 others were wounded, some of whom died later of gangrene, for the British were using dum-dum bullets, which inflicted terrible exit wounds. Ironically, with the threat from the Khalifa finally broken, the British began shooting each other in a series of unfortunate incidents. Orders had been given to shoot any tribesmen under arms or any wounded who looked dangerous, so the carefree soldiers sprayed bullets all over the place. The gunboats joined in and one shell ricocheted past Kitchener, and killed the *Times* war correspondent, Hubert Howard. And so Kitchener's luck held. Had two men performed differently that day Kitchener's whole career might have been changed. Had the Khalifa shown even the vaguest idea of how to control an army, or had Hector MacDonald shown the same sort of lunacy as Colonel Martin of the 21st Lancers, Kitchener might have lost the battle of Omdurman or at least suffered significant casualties. But Kitchener *was* lucky, and his military reputation, based as it was on the straightforward slaughter of medieval warriors by modern weapons, became inflated out of all proportion. Even his disastrous performance two years later in South Africa – at Paardeberg (see p. 148) – was not enough to stop his inexorable climb to the top of the British Army.

'How very amusing! Actually attacking our camp! Most amusing!'

Lieutenant-Colonel Henry Crealock, on receiving a report that the Zulus were attacking the British camp at Isandhlwana, 22 January 1879

This ludicrous comment by one of Lord Chelmsford's senior officers reveals the extent to which the British underestimated the threat from the Zulus in 1879. The result, of course, was the destruction of the British camp and one of Britain's most costly colonial defeats.

THE CENTRAL AFRICAN MISSION (1898)

The French army in the Sudan was a law unto itself. With scant supervision from Paris its officers saw it as a fast track to the top. Nobody cared to check very carefully how an officer carried out his orders, if indeed he was ever issued with any. As Napoleon had often said, you cannot make an omelette without cracking eggs. But sometimes officers went too far: they cracked the bowl as well. And when that happened the name of France itself was dragged in the dust.

Captain Paul Voulet – intelligent, well-educated and cruel – and Lieutenant Charles Chanoine – smooth, sophisticated and the son of a general – were the two officers chosen to lead the Central African Mission in 1898. Their 'mission' was rather vague and their orders virtually non-existent. If they could bring the territory east of Chad under French protection nobody was going to enquire too closely how they did it: provided, of course, that they kept everything under wraps. In army circles both men were known as ruthless killers who, the previous year, had marched through Senegal, laying waste to the countryside, burning villages and executing the natives. But they got the job done. Enough said: they were the men for the Central African Mission. Their orders were vague: 'I will neither pretend to indicate to you which route you must follow, nor the way to conduct yourselves with the native chiefs and the native populations.' This was from the Colonial Ministry in Paris. If ever a government was asking for trouble it was the French one in 1898. They were giving *carte blanche* to two known psychopaths in uniform. What followed – however horrific – was the fault of the French government and brought shame on France and the French people.

The 'mission' set off in November 1898. With Voulet and Chanoine went artillery expert Lieutenant Paul Joalland, Lieutenant Louis Peteau, Marine Lieutenant Marc Pallier and the feeble Dr. Henric as medical officer. Henric had been on the Senegal expedition with Voulet and knew exactly what to expect. In addition to three European NCOs, they took with them 50 Senegalese riflemen, 20 Sudanese horsemen and 30 locals to act as interpreters. The bulk of the fighting force was made up of 400 'auxiliary' troops, who were local thugs recruited by Voulet as bounty killers. The mission was extremely well armed, with artillery, machine guns, hundreds of rifles and millions of rounds of ammunition. In addition – although the French army had often encouraged its soldiers to 'go native' when the need arose – they had with them enough wine and champagne to maintain at least the semblance of civilization. But in their behaviour on this expedition they were to relinquish every hint of civilized values and rampage like crazed beasts. Their method of encouraging chieftains to submit to French protection was a simple one. They let loose their auxiliaries to slaughter whole villages in order to spread the fear of their coming, which quickly encouraged chieftains to seek their 'protection'.

The mission set out from Koulikoro on the Niger River. Chanoine took most of the party overland, across the 600-mile bend of the river, while Voulet and the rest of the men and supplies travelled downriver in canoes. At Timbuktu, Voulet spent a couple of days visiting the commander there, Lieutenant-Colonel Jean François Klobb. Klobb did not trust Voulet or his mission and anticipated serious trouble ahead. In Klobb's view, Voulet already had the glint of madness in his eyes.

Meanwhile, Chanoine and his black mistress, who ruled the men like a Cleopatra and had herself carried throughout, were already finding it difficult to feed such a large party. Once they realized that local villagers were unwilling or unable to supply them they abandoned all restraint and descended on the villages like a cloud of locusts, consuming everything. Before they linked up with Voulet again 148 porters died of starvation and dysentery. Those who tried to run away were ruthlessly shot down by Chanoine's Senegalese tirailleurs.

The two parts of the mission rendezvoused at Say on the Niger in January 1899. Here Lieutenant Peteau told Voulet that he had had enough and was leaving the mission. Voulet's response was to dismiss him formally for incompetence. This was a mistake that Voulet was to live to regret. Peteau at once wrote in disgust to his fiancée in Paris, a certain Mlle Lydia de Corvin, describing his experiences on the mission. Appalled by what she read, the young lady sent the letter to her local deputy, who promptly passed it on to the colonial minister. Soon questions were being asked in the French cabinet. On 16 April the minister contacted the Governor-General of the Sudan, Colonel Vimard, and told him to relieve Voulet and Chanoine immediately.

What they had learned from Peteau's letter were the details of Chanoine's overland trip to Say. Peteau wrote that Chanoine had shot twelve porters for trying to escape and had shackled hundreds of others by the neck in groups of five. When anybody broke down through illness or exhaustion they were decapitated on the spot. In one village Chanoine ordered 20 women and children to be bayoneted, while in another over a hundred were massacred. Black NCOs were sent out to round up porters; as evidence of their thoroughness they brought back the heads of those who resisted, which Chanoine paraded on pikes. The mere sight of the tricolour flag in the distance caused whole villages to flee. Voulet and his companion maintained a savage discipline within their own force, killing anyone who disobeyed an order, and flogging the porters like slaves. One soldier who lost some bullets on the way was summarily executed by Voulet.

Something had to be done and done fast, but how could the French authorities reach the mission, which was by now far from any French post? Eventually, in April 1899, the decision was reached to send Lieutenant-Colonel Klobb from Timbuktu. Just as Klobb was preparing to return to France on leave he received the order to bring in the renegade officers. With just 50 tirailleurs and a young Parisian lieutenant, Octave Meynier, Klobb set off in pursuit of the mission. He did not need guides – the vultures showed him the way.

Chanoine, in the meantime, was succumbing more and more to paranoia. He believed their difficulties were all part of a government plot to sabotage the mission. Surely it would be better to break with France and branch out on their own. Voulet agreed. He told Chanoine that he was no longer a junior officer; he was a warlord, well-armed and with an army of obedient killers. Let the French try to take him back. He would fight them as well.

As the two men grew more crazed they became even more ruthless. Guides who failed to find food were hanged from trees, low enough for the hyenas to eat their legs, 'while the rest was left for the vultures'. Anyone who could not keep up was shot out of hand. When one villager tried to resist and killed two tirailleurs Voulet ordered 150 women and children to be slaughtered. And all the time, following this grisly trail of burned villages and charred corpses came Klobb, slowly but relentlessly closing in on the mission. As he advanced he passed trees from which hung the bodies of women, and villages where children had been burned in the cooking fires. It was like a descent into Dante's inferno. Klobb could hardly believe that such atrocities were the work of French officers.

On 10 July Klobb reached the village of Damangara and heard from its frightened inhabitants that Voulet was just a few hours' march ahead. He sent one of his NCOs with a letter to Voulet, ordering him to relinquish his command at once. Voulet and Chanoine now faced the crisis they had been expecting for some time. The only way out for them was to destroy Klobb and his whole force, but they must not let the other Europeans see them do it. They dispersed their fellow officers and European NCOs to carry out raids or locate wells, while they took care of Klobb. As soon as the white witnesses had gone they sent Klobb's NCO back with a reply, telling him that they were short of water and would meet him at the next village.

Klobb was alarmed at what he read in Voulet's letter. The man was clearly mad. He had no intention of returning to France or facing his accusers. He even dared to threaten his superior officer with violence. But Klobb knew his duty. The next day he led his column forward until he was spotted by one of Voulet's lookouts. Voulet assembled his men in line of battle and sent a final warning to Klobb: keep away or face the consequences. But Klobb was indomitable. He did not believe that Voulet would dare to shoot him in front of European witnesses. Ordering Meynier not to open fire under any circumstances Klobb went forward alone towards the line of tirailleurs, standing in the midday heat with their rifles pointing directly at him. Voulet called to Klobb to go back but he refused. He began to call to Voulet's men, reminding them of their duty. That was the final straw. Voulet gave the order to open fire and at a range of a hundred yards the tirailleurs fired, injuring Klobb, and knocking him to the ground. Propping himself up on one elbow Klobb called back to Meynier reminding him not to fire. As he spoke a second volley rang out, killing him and wounding Meynier. At the sight of this Klobb's soldiers broke and ran, leaving Voulet master of the field.

Voulet now walked over and began haranguing

Meynier with a list of his 'imagined' grievances. Klobb was his enemy – had always been his enemy – and was trying to steal his supplies and ruin the mission. Meynier could do little but listen as Voulet raved on about the crimes that France had committed against him and how he had renounced his country and was now a 'black chief'. He told the injured man, 'I am going to create a strong empire, impregnable, which I shall surround with an enormous bush without water. To take me, it will require 10,000 men and 20 million francs . . . If I were in Paris I would be master of France.' Meynier wisely kept his thoughts to himself. Next Voulet told his black soldiers that they had abandoned France and that they were deserters whom the French would shoot. They must follow him as his warriors. But they had noticed that none of the other European officers seemed intent on following Voulet and Chanoine. Dissent broke out in the camp and Voulet threatened to shoot anyone who tried to leave. But by now discipline had broken down and now it was every man for himself. Voulet and Chanoine lined up the tirailleurs and began shooting at them. The Senegalese fired back killing Chanoine instantly but Voulet escaped into the darkness, finding refuge with some villagers. The next morning he rode back into the camp, but a sentry challenged him and would not let him through. Voulet shot at the man, who responded by shooting him dead.

The death of Voulet broke the spell that had held the entire mission in thrall. Like the inhabitants of the castle in 'Sleeping Beauty', the European officers suddenly awoke from their bad dream and decided to continue the mission as if nothing had happened. They were French officers doing their duty, and they could not be blamed for that. Voulet had killed Klobb, not them. It had all been out of their control. Lieutenant Joalland took command and the mission passed on its way.

When the news of Klobb's murder and the atrocities committed by Voulet and Chanoine reached France there was a storm of indignation in the national press. The nation was ashamed and the army was in disgrace. France's claim to have a 'civilizing mission' in Africa became the subject for cheap jibes in foreign capitals. The army defended itself by saying that the dreadful heat of the Sudan had affected the minds of Voulet and Chanoine – but everybody knew better. How could men who were known to have committed atrocities in Senegal have been given command of another expedition? The fault lay with a system that encouraged men to use the Sudan as a way of bettering their career without feeling any obligation towards the people who suffered at their hands. The desire for promotion, fame and riches brutalized the French soldiers in the Sudan; there were many Voulets looking for a chance to 'go native' in the interior and emerge with a colonel's rank and pension. If the people in the interior of Africa suffered, who would ever know?

The army smoothed its ruffled feathers and rationalized the problem. Africans had always suffered. If it was not French tirailleurs killing them it would be Arab slave traders or local warlords. Tragically there was no Livingstone in the Sudan to give them the lie.

Major Francis Clery: *'What is to be done on this report?'*
Lord Chelmsford: *'There's nothing to be done on that.'*

Clery had just received the following message from Colonel Pulleine at the camp of Isandhlwana: 'Report just come in that the Zulus are advancing in force from left front of camp, 8.05 am.'

Chelmsford, underestimating the danger from the Zulus and refusing to form a laager, as the Boers had advised, had divided his forces and led half of them out on reconnaissance. He could not believe that a savage and primitive foe like the Zulus would have the effrontery to attack a camp held by British regulars. His education was to be completed at the 'school of hard knocks'.

THE BATTLE OF PAARDEBERG (1900)

At Omdurman in 1898 Lord Kitchener had revealed more than a hint of panic, though it had been overlooked in the euphoria of a great victory. Yet, two years later, in South Africa, his extraordinary conduct at the battle of Paardeberg was to earn him a title that would not soon be forgotten – 'King Chaos'. It was not a nickname delivered with any kind intentions. Kitchener had wasted lives through an incompetent handling of the British soldiers under his command. They were unlikely to forget or forgive him.

In February 1900 the main Boer army of 4,000 men under General Cronje had established a laager on the Modder River near Paardeburg. Facing them was a British army of 15,000 men commanded, in the absence of Lord Roberts who was indisposed with a chill, by Major-General Lord Kitchener, who had been preferred by Roberts to the more experienced and more cautious Lieutenant-General Kelly-Kenny. While Kelly-Kenny proposed to bombard the Boers into surrender, Kitchener overruled him, deciding to take the laager by infantry assault. To Kelly-Kenny's horror and disgust Kitchener was proposing to ford a river and launch a frontal attack on well-entrenched Boer riflemen. The British had been through all this before at the battles of the Modder River and Colenso (see *The Guinness Book of Military Blunders,* pp. 37–8). Everyone knew it was a formula for absolute disaster. Yet Kitchener was deaf to complaints. Buller had failed at Colenso because Buller was a fool. Kitchener boasted that he would show how it should be done. As he told one officer, 'We'll be in the laager by half-past ten!'

The problem was that if Buller was a fool Kitchener was a 'loner', who did not know how to delegate. He had great faith in himself but virtually none in anybody else. He was also unable to issue coherent orders, always knowing what he wanted to do but being quite unable to explain it to others. He hated writing things down. He never seemed to have time; he was always rushing off somewhere before he had explained himself fully. The fog of war is present in every battle but at Paardeberg Kitchener seemed to produce most of it himself. Yet if he issued chaotic orders it must not be forgotten that Roberts had sent him to fight a battle without any staff. Not that this can have bothered Kitchener very much. Kelly-Kenny later complained that Kitchener sent only verbal messages. 'Kitchener of Khartoum' soon became 'Kitchener of Chaos' in the view of many senior officers.

For two miles the British infantry advanced across an open plain in full view of the Boers, who were hidden behind wagons and up trees across the river. As the British closed in on the laager they were hit by sheets of rifle fire from the Boers and were forced to lie down for cover. While the first wave of infantry took cover, more and more of Kelly-Kenny's infantry poured across the plain urged on by Kitchener. Nobody got closer than 200 yards from the riverbank. Casualties were extremely heavy.

Meanwhile, the Highland Brigade were sent to join the rush for the river. Their commander, General Colville, found this most mysterious as he had not ordered them to move. Naturally it had been Kitchener, riding up, shouting orders to junior officers, and then flying off again like the wind. All Colville could do was watch as his men wheeled round and joined the frontal attack, only to be scythed down by the Boer rifle fire. To those who had been with Methuen at the Modder River in November it was a case of *déja vu*. British soldiers, on their faces with the sun burning their backs and Boer bullets whizzing over their heads.

Kitchener was not pleased. Colville's men had not tried hard enough. They must go again. Were there any fresh troops? Colville told him that there was just a half battalion of men guarding the baggage. Kitchener ordered them to attack straightaway, ford the river and take the laager. Colville's jaw dropped. Nobody had got anywhere near the river yet. He told the men to eat lunch; at least let them die with full stomachs. But Kitchener rushed back to cancel their lunch and tell them to attack forthwith. Kitchener next told Kelly-Kenny that his men had not done well enough either. They must try again. The two generals were at daggers drawn. Only by a great effort of will did Kelly-Kenny prevent Kitchener squandering the rest of his men in the frontal assault. Both the division's brigadiers were wounded at the field hospital and he was himself suffering from dysentery. He felt that Kitchener would be the death of him – and the army.

Lord Kitchener of Khartoum earned a high reputation for his defeat of the Mahdists at Omdurman in 1898. But two years later – during the Second Boer War – he earned the title of 'King Chaos' for his mishandling of the battle of Paardeberg.

Kitchener now rode over to order Colonel Hannay to attack the right flank of the laager and take it 'at all costs'. Hannay's mounted infantry were told to gallop right up to the laager and fire into it. This was a callous and reckless order. It was nothing less than a death sentence and it was absolutely unnecessary. Nobody else on the battlefield had been told to take the laager 'at all costs': the imperative request seemed reserved for Hannay alone. Hannay received his orders – seemingly those of a madman – with an air of resignation. Gathering just 50 men around him he charged the laager and was shot down, still 200 yards from the Boer trenches, but the closest that any British soldier got that day. Hannay's death was somehow a protest against the absurd martyrdom that Kitchener was inflicting on his army. Even the Boers were impressed. As one of them said, 'We didn't want to shoot a man so blatantly inviting death and so defenceless. But we were afraid that he was running berserk or would suddenly wheel away to give information about us. We had no choice.' The spot where he fell is today marked by a monument.

On the far right of the Boer position Brigadier Smith-Dorrien (later to command a corps of the BEF in France in 1914) and his men had waded across the river and had sat waiting for orders from Kitchener. For nine hours they waited and heard nothing but the fire of Boer mausers and the explosions caused by the British artillery. Had they been forgotten? Smith-Dorrien was experiencing his own 'fog of war'. Without contact with any other British contingent, he had no idea what was going on. Another hour passed and then Smith Dorrien was amazed to see the right wing of his brigade suddenly rise up and rush to the attack. Amazing, for he had not issued a single order. Kitchener was up to his tricks again. While Smith-Dorrien watched, his men were shot to pieces by the Boers, none getting closer than 300 yards from the laager.

Sunset came. Kitchener could see that nothing more could be done that day and a merciful silence descended on the battlefield. The British had suffered their biggest defeat of the entire Boer War, with 24 officers and 279 men killed and a further 900 officers and men wounded. Reporting to Roberts Kitchener gave the impression that little less than a victory had been achieved. Roberts was not so sure. Kitchener wanted more of the same the next day. Fortunately, Roberts was feeling better and resumed command, thereby saving the lives of the hundreds of British soldiers Kitchener presumably had it in mind to squander on the morrow. Even so, once Roberts reached Paardeberg he did not like what he saw and considered retiring. Fortunately the much tried Kelly-Kenny was adamant that there should be no retreat. The truth was that Cronje was in an impossible position if only the British commanders would realize it. He could be bombarded into submission if Kitchener would stop scattering British soldiers' lives as if they were of no account. Under no circumstances must the Boers be allowed to get away from under the British guns. Kelly-Kenny's resistance saved Roberts from making the biggest blunder of the entire war and the worst mistake of his whole career (we will draw a veil over his choice of Kitchener to command the battle of Paardeberg).

Just two days later – ironically on the 19th anniversary of the battle of Majuba Hill – Cronje, with 4,069 Boer soldiers, surrendered the laager. Roberts had triumphed over himself; Kelly-Kenny had triumphed over Roberts; Kitchener had avoided the court-martial that his insane generalship richly deserved. Had he commanded in France in 1914, rather than served as Minister of War, one shivers to think what ghastly exhibitions of incompetence could have occurred. Historians might even have written of the relief felt in the army at his replacement by John French or Douglas Haig sometime in 1915.

'Get up, you lumps, and behave like Englishmen.'

A British officer addressing Australian troops of the Victoria Rifles during the Boer War

Not, one would suggest, the way to get the best out of Australian troops. Even as late as the Second World War British officers failed to understand the relaxed and apparently casual Aussie soldier's approach to discipline and army 'bull'.

THE BATTLE OF DUL MADOBA (1913)

The 'Mad Mullah' loomed large in the stories of imperial 'derring-do' in the last years of the 19th century. Inspiring his fanatical hordes of Dervish warriors with tales of paradise to come, the Mullah soaked the desert sands of the Horn of Africa with the blood of the infidels. His cruelty made him a name for English nannies to invoke when young masters would not eat their greens or go to bed on time. The Mullah, a master of disguise, was the sworn enemy of a regiment of British heroes in 'Boy's Own' tales. He was both man and legend, and it was sometimes difficult to separate truth from fiction in dealing with him.

Just a year before the holocaust of 1914, Somalia was still a land of mystery and intrigue, a fitting setting for novels by Rider Haggard or G.A. Henty. And from their pages stepped the man who would bring the 'Mad Mullah' to account or die in the attempt: tall, handsome Old-Marlburian Richard Corfield, Britain's answer to 'Indiana Jones'. Accompanied by the Camel Corps Corfield would cross deserts to rescue kidnapped English ladies from the tents of the foul Dervishes; he would swim crocodile-infested rivers to wrestle single-handed with giant Negro slaves who guarded the Mullah's secret hideout. But he was also capable of leading his men into a trap because he was too hot-headed to think clearly.

Acting-Commissioner G.F. Archer was not a romantic. He was genuinely alarmed at the prospect of letting the impetuous Captain Richard Corfield loose to make mischief. He knew how eager Corfield was to reimpose Britain's authority on the Somali tribes. It would make a good plot for a novel. But Archer had no time for novels. It was his responsibility to prevent trouble in this obscure corner of Britain's world empire. Nobody was going to thank him for a military disaster. He was not in the 'Chinese' Gordon mould, and he did not intend to let Corfield make any more trouble for him. There were already too many newspapermen trying to get stories about the 'Mad Mullah' for his liking. It was in this state of mind that Archer spoke to Corfield in June 1913, warning him in no uncertain terms that if he took the Camel Corps out against the Dervishes there must be no disasters. Corfield must confine his activity to the immediate vicinity of Burao. If he found himself in trouble with the Dervishes he must fall back at once. Yet even as Archer spoke he noticed that Corfield was already looking away towards the sands stretching for hundreds of miles towards the Red Sea.

Corfield was chafing at the bit, pressing to advance far beyond Burao to protect the tribes from attacks by the Mullah's men. He did not need some Colonial Office pen-pusher telling him what was needed. He tried to enlist the help of some village elders to convince Archer to let the Camel Corps take action against the Mullah, but the Commissioner was adamant. He would permit a reconnaissance only. To keep a check on Corfield and, if need be, restrain him, Archer sent Captain G.H. Summers – tall, upright and incredibly sensible.

On 8 August the Camel Corps left Burao, with Corfield riding out at the front and with his officers and 110 mounted soldiers behind, pulling the Gatling gun on which all their hopes rested. As they passed through the villages they were joined by 600 spearmen and 2,000 riflemen from pro-British tribes. It was a strong force and it should have been enough to keep Corfield out of trouble. But trouble and Corfield had a way of meeting each other head on. As the camel column travelled south, Corfield received news that a large Dervish force led by the Mullah's brother was heading towards the town of Idoweina driving a huge flock of animals that they had stolen from the tribesmen. This was just the chance that Corfield had been waiting for. He sent out scouts to confirm the sighting and pressed on towards Idoweina, hoping to cross the Dervish line of march.

By the time darkness fell Corfield and his column were about ten miles from Idoweina. Their scouts had confirmed the first sighting and added that there were about 2,000 Dervish warriors approaching Idoweina from the east. Corfield pressed on a little before setting up camp just four miles from Idoweina, well within sight of what he could see from the campfires was the Dervish camp. Alive to the danger of a night attack, he surrounded his camp with bushes and bracken and set up the Gatling gun in the middle, ready for immediate use.

But Corfield could not sleep. With the enemy so close he was eager to strike a blow at them. He discussed with Summers the idea of making a night attack but Summers managed to persuade him to wait, reminding him that the Camel Corps had no experience of night fighting. Yet Corfield was not to be held in check for long. He told Summers that he would wait until dawn and no longer. Then he would attack. As the sun rose over the desert scrubland the Camel Corps marched out of camp, leading their camels and with their Gatling gun in front. From the clouds of dust they could see that the Dervishes were already on the move, in a line parallel to Corfield's march. Through his field glasses Summers reported seeing a group of Dervish horsemen riding towards them. Corfield's men set up the Gatling and everyone stood at the ready to repel the riders. It was not until they were quite close that Summers was forced to admit that it was just a troop of ostriches. Corfield scowled; Summers was a fool.

Shortly afterwards there was a genuine sighting of Dervishes, and Corfield dismounted the corps and formed a square, ready to repel an immediate assault. But as none transpired he decided to move his men forward through the thick bush and onto open ground, where the Gatling's field of fire would be clearer. But Corfield had catastrophically miscalculated the speed of the Dervish advance. As the Camel Corps struggled through the dense bush, they were hit by the Dervish horsemen before they could form a square. The result was disastrous. Corfield's line wavered and broke, and soon the Dervishes were screaming and hacking at the troopers from all sides. Corfield's native allies now panicked and fled, carrying much of his ammunition with them. The Dervishes attacked in waves, each of which broke over Corfield's shattered command, thinning his ranks at every charge. Matters took a further turn for the worse when 25 of the Camel Corps broke ranks and fled, while the vital Gatling gun was struck an unlucky blow by an enemy bullet and rendered inoperative. Corfield remembered his Newbolt – 'the square that had broken . . . the Gatling gun jammed . . .' What was that final line? Corfield ran to the Gatling and tried to get it firing again but as he knelt by it he was shot once, twice . . . Yes, that was it, 'Play up, play up, and play the game'. He was hit by another bullet through the head and killed.

Summers now stepped into Corfield's shoes. He formed the remaining Camel Corps troopers into a circle, sheltering behind a ring of dead camels and ponies. He too was wounded. With the Camel Corps at its last extremity, ringed round by thousands of screaming Dervishes, an extraordinary event occurred, too strange even for a Haggard or a Henty. The mounted Dervishes suddenly wheeled away and rode into the distance, followed by their hordes of spearmen, leaving 600 of their number dead on the ground. Why they suddenly disengaged nobody knows. With the Camel Corps now reduced to just 40 men, and with the sorely wounded Captain Summers in command, escape had seemed impossible. Yet it had happened. Had the Dervishes perhaps seen something? Possibly the shadowy figure of Corfield himself, standing, revolver in hand, within the circle of troopers? Or had it just been a trick of the light?

In many ways it had been a pointless engagement. Corfield's impetuosity had got him into exactly the kind of trouble that Archer had feared. Corfield had gone looking for trouble and he had found it. Tactically he had allowed himself to be caught in thick bush which he must have known would prove fatal. Half a century of British experience in Egypt, Sudan and Somalia had shown the vital importance of preserving the square against desert warriors. If this was broken – as once before at Abu Klea – the Muslim swordsmen were a match for even British regulars with their bayonets. The fanaticism of the Dervishes made them immune to fear, and the fire

'The attack on the Somme bore out the conclusions of the British Higher Command, and amply justified the tactical methods employed.'

Colonel Boraston, biographer of Haig, 1919

This extract from an early biography of Haig shows how completely the 'establishment' view of the Somme campaign prevailed at that time.

of a Gatling gun and concentrated rifle fire was usually needed to halt them. Courageous though he had been, Corfield's headstrong nature was his undoing. He should have exploited the strengths of the Camel Corps and concentrated on hit-and-run attacks on the larger Dervish force to regain the animals looted from the pro-British villages. Tragically, the plucky Corfield had played into the hands of a cruel enemy.

THE BATTLE OF THE SOMME (1 JULY 1916)

General Sir Douglas Haig has become so identified in the public mind with the carnage on the Somme in 1916 that the role of General Sir Henry Rawlinson, commander of the British Fourth Army, has been largely overlooked. But Rawlinson played a greater part in the planning and execution of the great British offensive than Haig, and some of the failings which turned the operation into a bloodbath were his. On 1 July 1916 – the first day of the battle and the blackest day in the history of the British Army – it was Rawlinson's lack of moral courage that resulted in thousands of British and imperial soldiers being sent to certain death. The 34th Division – facing the wire opposite La Boiselle – was sacrificed because Rawlinson lacked the courage to tell Haig that his artillery had failed and that as a result the whole operation had been compromised. The wire facing 34th Division had not been adequately cut and thousands of men would have to advance against unbroken wire, in places 20 feet thick, with no chance whatsoever of reaching the German trenches. All that they could do was to 'face to the front' as the German machine guns mowed them down. In all, 6,380 men were shot down from this one division, nearly 75 per cent of its infantry and the highest figure of any division attacking that day. Rawlinson must have known that this would happen and that he could prevent it. To have done so, however, would have called into question the whole strategy on which the British plan was based, and it would have taken a braver man than Rawlinson to have done that. So he let the attack go ahead and stiffened his resolve for when the casualty figures came in, no doubt reassuring himself that in the context of an attritional struggle one day's figures were statistically insignificant. He may even have asked himself, who would remember one day's fighting in a battle lasting five months? But when the figures did come in Rawlinson – now thoroughly alarmed for his reputation – ordered his staff to destroy all the notes he had issued as a guide to the officers leading the attack. Staff at Fourth Army later doctored the account that went into the Army War Diary to remove any evidence that might suggest that Rawlinson had blundered.

Yet Rawlinson and Haig were mistaken if they thought that it could all be swept under the carpet, for the casualty figures of 1 July 1916 have never been forgotten or forgiven. Though the army recovered from its losses and fought on for a further five months, on 1 July 1916, the nation suffered a psychological blow which left scars that have never healed.

By the end of 1915 it was apparent to the British government that there was no prospect of the war against the Central Powers being won without Britain assuming a much greater share of the burden of fighting. French losses in their spring and autumn offensives of 1915 had stretched their manpower to its limits. It was obvious that the vast numbers of Britons who had responded to Lord Kitchener's appeal for volunteers in 1914 would now need to be used in France. British politicians did not reach this conclusion without a struggle. During 1915 attempts had been made to find an alternative winning solution by undertaking campaigns in Gallipoli and Salonika. But in neither of these peripheral areas was Britain able to force the Germans to withdraw troops from the main battleground in Belgium and France. Thus when the Allies met at Chantilly to coordinate their strategy for 1916, Britain was obliged to commit herself to a summer offensive in France, in conjunction with a major French offensive there.

On 7 April 1916 the British government authorized their new commander-in-chief, Sir Douglas Haig, to concert an offensive with the French. The battlefield was to be the Somme region – in the words of Sir Henry Rawlinson, 'The country resembled Salisbury Plain, with large open rolling features, and any number of partridges which we are not allowed to shoot.' In fact, apart from the partridge-shooting, the Somme was a very bad choice. The German front lines there enjoyed the

General Sir Henry Rawlinson (left) and Field-Marshal Sir Douglas Haig (right) on the steps of the British headquarters at Querrieu in 1916. The first day of their offensive on the Somme saw nearly 60,000 British casualties.

advantage of the high ground, so that the waves of advancing British soldiers would face a climb towards the strategically vital Pozières ridge. In addition, the chalky subsoil had allowed the Germans – unchallenged in the area since October 1914 – to construct intricate and effective underground defences, resistant to artillery barrage. Nevertheless, the Somme was the area chosen by French commander Marshal Joseph Joffre, and Haig felt obliged to comply. At first, Joffre had planned an operation in which equal numbers of French and British troops would be involved, but after 21 February – when the Germans began their massive assault on Verdun – the French commitment to the Somme diminished. By May Joffre was only able to offer thirteen divisions to support the 20 or so British divisions earmarked for the assault. By the late spring of 1916 British strategic options had narrowed. With the pressure around Verdun threatening to break the French army, Haig had little alternative but to attempt to lift the siege by diverting German troops to the Somme.

But what sort of offensive were the British planning? Experience at Loos and Neuve Chapelle in 1915 had demonstrated that unsupported infantry and cavalry had no chance in the contested zone while enemy firepower remained unsuppressed. Moreover, territorial gains were only temporary if the Germans were able to feed in reserves at threatened points of their line and to stage effective counter-attacks. The only solution seemed to lie in firepower. On the Somme it would be necessary to overwhelm the German defences with an artillery bombardment of unprecedented weight and ferocity. Once the artillery had succeeded in demolishing the enemy it would be a relatively simple task for the infantry to occupy the ground won. It all sounded very straightforward.

Haig had chosen Fourth Army commander, Sir Henry Rawlinson, to spearhead the offensive on the Somme. With the French on their right, Rawlinson's army was to attack along an 18-mile front between Gommecourt in the north and Montauban in the south. Half a million men under his command would prepare the most prodigious military operation ever undertaken by a British army. In some ways Rawlinson – an efficient administrator and an effective infantry general – was a good choice, yet in one vital area he did not see eye-to-eye with Haig about the forthcoming battle. And in this disagreement between commander-in-chief and army commander lay the seeds of disaster.

A staggering amount of preparation was needed behind the British lines in the weeks before the operation was to take place. A whole new town had to be built to accommodate the assault troops, to feed and equip them, and to provide them with medical services and entertainment. Hundreds of thousands of horses and transport vehicles passed endlessly up the newly constructed roads and railways, while field guns, howitzers and mortars in unprecedented quantities were moved into position and hidden from prying German eyes. Yet how could all this preparation take place in secrecy? In the first place, the Royal Flying Corps drafted aircraft into the Somme region and established air supremacy over the British lines to keep German reconnaissance aircraft at a distance. In France every effort was made to keep the assault secret. But in Britain no such secrecy prevailed. All efforts to keep the Germans guessing were ruined when British newspapers reported that munitions workers had had their Whitsun leave cancelled and been put on round-the-clock rotas. Nothing could be clearer to the Germans than that the British were planning a major offensive. But where would it come? In Flanders, where the bulk of the British troops were based, or further south? It did not require a genius to conclude that British aerial activity on the Somme was an attempt to conceal major troop concentrations there. As a result, the Germans were quite prepared for the British attack whenever it might come. They concluded that they could rely on the British generals to give them ample warning. In the event, Rawlinson's seven-day artillery barrage was as clear a signal as anyone might need.

Infantry training was taken very seriously by the officers, but probably raised more than a few laughs from Tommies with a feel for black comedy. Behind the lines a huge area of sandy, dusty soil was laid out with tapes. The men then went 'over the top' and captured the 'tapes' – it was all bloodless and purely symbolic. As the Tommies went into action they had to imagine gas (a quick drag on a fag and a few coughs should take care of that), and clamber over imaginary barbed wire ('No cheating there! Get that leg over!'). To make matters worse everyone had to walk as if they were fully loaded with 70 pounds of kit, wearing gas masks, carrying spades, pigeons, rolls of barbed wire and extra bombs. Such training was considered necessary by the officers, who felt that the men could not be trusted to do anything as simple as run in a straight line or take cover in shell holes.

While the soldiers toiled, their commanders were at odds with each other as to what exactly was the aim of the offensive. Haig was still thinking in terms of a 'decisive battle', which would break the German lines and enable him to use his massed cavalry corps to burst through into open ground. Thus Haig hoped to punch a hole and pour through it in a grand Napoleonic sweep. Unfortunately for him, and perhaps for his men, Rawlinson did not agree with him. He had lost hope of a breakthrough and preferred to think in terms of a 'bite and hold' operation. He rejected Haig's belief in 'going for the big thing' and, instead, aimed to deliver sharp blows against limited targets, that would exhaust the German reserves as they tried to repair holes at various points in their line. So, in spite of Haig's own belief that the Fourth Army was aiming for a decisive breakthrough of both the first and second German defensive lines on 1 July, it was in fact merely proposing to take the first positions, before consolidating and moving up the artillery to support further advances. How Haig and Rawlinson could have arrived at such different interpretations of their task after such lengthy discussions is difficult to understand.

As alarming – and even more disastrous in its human consequences – was Rawlinson's application of infantry tactics. As a professional soldier he seemed to have little faith in the men who would compose the assault force on 1 July. Sixty per cent of them would be men from Kitchener's 'New Army' – volunteers and ex-civilians – who in his eyes could not be trusted to do anything properly. As a result, whereas the French infantry on the British right advanced across no-man's-land in small, tight groups, rushing from one piece of dead ground to another, covered all the time by other groups behind them, the British soldiers – weighed down by 70 pounds of equipment – were ordered to advance upright, a yard or two apart from their neighbours and at a walking pace, to prevent them panicking and diving for cover. A training memorandum issued just three weeks before the attack by Haig's Chief of Staff, Sir Lancelot Kiggell, ordered the attacking infantry to advance in four rows. Kiggell warned that heavy casualties could be expected and did what he could to guarantee them by having the British advance like tin soldiers waiting to be swept away by a child's arm. The German defenders were later to write in astonishment about the slow, steady march of the British who, had they come at a rush, would certainly have succeeded in capturing many more trenches. Paul Scheyt wrote:

> The English came walking, as though they were going to the theatre or as though they were on a parade ground. We felt they were mad. Our orders were given in complete calm and every man took careful aim to avoid wasting ammunition.

In the words of Musketier Karl Blenk:

> When the English started advancing we were very worried; they looked as though they must overrun our trenches. We were very surprised to see them walking, we had never seen that before. I could see them everywhere: there were hundreds. The officers were in front. I noticed one of them walking calmly, carrying a walking stick. When we started firing, we just had to load and reload. They went down in their hundreds. You didn't have to aim, we just fired into them. If only they had run they would have overwhelmed us.'

A French artillery observer commented, 'I thought of the Crimea today, and of what the French said in the Crimea about the Charge of the Light Brigade.'

Rawlinson's tactics were a formula for disaster. How could such absurd instructions be given to soldiers marching in some places as far as a thousand yards into barbed wire, all the time under raking fire from machine guns? In his own defence, Rawlinson would have insisted that this scenario was wrong, for the simple reason that there would not be any machine guns, or barbed wire, or even live Germans. The British infantry was advancing to occupy ground won for it by the devastating weight of the British bombardment. And in this assumption lay the core of the disaster that was to strike Fourth Army on 1 July 1916.

Rawlinson's confidence in the power and effectiveness of his artillery was absolute. As he told his officers, 'nothing could exist at the conclusion of the bombardment in the area covered by it.' He was using 1,437 guns on a 15-mile front and his guns would achieve three things:

(1) They would suppress German artillery fire when the assault began.

(2) They would destroy the German barbed wire – even though some of it was so thick that light could scarcely pass through its close-meshed coils.

(3) It would kill all the German soldiers in their trenches, dug-outs and bunkers, so that there would be no one to scythe down the walking waves of Fourth Army.

But just how effective would the British bombardment be? Previous barrages had only ever succeeded where the Germans were taken by surprise or had poor bunkers in unsuitable terrain. Was this the case on the Somme? The answer is 'No'. In spite of every effort to keep them in the dark about British intentions, the Germans knew where and when the British attack would take place. Even worse, the Germans had been in their Somme positions for two years and had used the time to build the best and deepest defensive positions on the whole of the Western Front. In places the British would meet four separate trench systems and might have to cross twelve trenches before reaching open country. Concrete dug-outs 30 feet deep kept the soldiers in safety during the barrage, and barbed-wire entanglements of awesome efficiency lined the forward trenches. This was what Rawlinson was dismissing so airily when he spoke of the irresistible power of his big guns. And his confidence in the guns was quite misplaced. Of the 1,437 artillery pieces available, only 467 were heavy guns and of those just 34 were of 9.2 inch calibre or more. In the event, just 30 tons of explosive were to fall on each mile of the German front – hardly impressive when the distinguished military historian John Keegan has suggested that such powerful defences would today warrant

A narrow escape for a British 'Tommy' during the Somme advance. Bandaged at a field dressing station, he brandishes the helmet that saved his life from German shrapnel.

several small nuclear warheads. Even worse was the nature of the shells that would be fired. Nearly two-thirds would be shrapnel, deadly to men in the open, but harmless to those in deep dug-outs. The shrapnel could cut the barbed wire if fired with precision, but there is no evidence that the British gunners had developed such skills at this stage. Of the 12,000 tons of explosive fired by British guns in the last week of June 1916, just 900 tons were of high explosive capable of destroying the deepest German defences. And when one considers that nearly one-third of the shells fired – many of low-quality American manufacture – failed to explode at all, one can see how disappointing was this supposedly all-destructive barrage. Nor was accuracy a virtue of the British artillery at this stage of the war. British shells could never be used less than 300 yards ahead of the British troops – but the front-line trenches were often not that far apart. As a result, much of the artillery fire missed the German front-lines, machine-gun positions and concrete pill-boxes. The British figure of a 300-yard safety limit should be contrasted with a French one of 60 yards and a Japanese one (in the Russo-Japanese War eleven years before) of 100 yards to illustrate how mistaken Rawlinson and Haig were to rely on the British gunners to provide a 'creeping barrage' for their infantry or to silence the German front lines and their wire defences. In simple terms the gunners were not up to the job.

Now what do you do if you are a commander who has staked everything on the success of an artillery bombardment which has clearly failed? Do you show the moral courage necessary to call off the attack and go back to the drawing-board? Or do you simply pretend that everything has happened as you predicted? Both Haig and Rawlinson are culpable in that, in spite of intelligence reports that spoke of unbroken barbed wire facing three of the five corps due to attack on 1 July, and of prisoners taken who clearly had suffered little from the intensive barrage, they went on blithely believing what they wanted to believe rather than the evidence of their own eyes. Haig even remarked fatuously, 'The barbed wire has never been so well cut, nor the Artillery preparations so thorough.' Haig clearly had no real idea of what was going on at the front. On the other hand, Rawlinson himself wrote, 'I am not quite satisfied that all the wire has been thoroughly well cut and in places the front trench is not as much knocked about as I should like to see in the photos. The bit in front of the 34th Division has been rather let off.' As we have seen, it was the 34th Division which was to suffer massacre on 1 July. And was forewarned forearmed? Not a bit of it – Rawlinson ordered the attack to proceed as if the barrage had been totally successful. The Sherwood Foresters were told:

> You will meet nothing but dead and wounded Germans. You will advance to Mouquet Farm and be there by 11 am. The field kitchens will follow you and give you a good meal.

The King's Own Yorkshire Light Infantry (who incidentally suffered 76 per cent casualties) were instructed:

> When you go over the top, you can slope arms, light up your pipes and cigarettes, And march all the way to Pozières before meeting any live Germans.

At 7.30 am on 1 July the first waves of 60,000 men went over the top and marched slowly towards the German lines. Some were led by officers who kicked footballs, others by men with walking sticks or umbrellas. Within 30 minutes half of them had become casualties. Of the 120,000 from 143 battalions who attacked that day nearly 60,000 casualties were suffered, including some 20,000 dead. It was the greatest loss ever suffered by the British Army and the heaviest by any army in a single day of the entire war. In fact, British battle casualties that day exceeded those from the entire Crimean War, Boer War and Korean War put together. The cause was simple. The German soldiers had survived the barrage in their deep concrete bunkers and, warned by the cessation of the bombardment ten minutes before zero-hour (another costly error by the British), they emerged from below ground and reached their machine guns before the slowly advancing British troops were even halfway across no-man's-land. In those areas where the British did succeed in taking German positions, they found them intact and even with the electric lights still working. So much for Rawlinson's crushing barrage.

The British troops encountered not just rifle and machine-gun fire, they also marched into the face of a counter-barrage from the German artillery, which had similarly survived Rawlinson's bombardment. Even where the wire had been broken the passages created served as death-traps, for the Germans had concentrated their firepower on these openings.

Elsewhere, the wire trapped thousands of men, equipped only with pitiful hand-cutters, who could find no way through and milled about like flocks of sheep until the machine guns scythed them down. One battalion of the Newfoundland Regiment suffered 91 per cent casualties on the wire.

The regiments which suffered worst were the 'Pals' Regiments', set up by Lord Kitchener. To encourage recruitment men had been allowed to form regiments with their friends from factories, unions, churches, sporting clubs, craft guilds, schools and many similar organizations. The entire first team from Heart of Midlothian Football Club was in action on 1 July. There were even 'Bantam' regiments of miners, set up because of the pit workers' small stature.

As wounded men began to stream back, anxious battalion commanders in the second wave telephoned for new orders but the answer was always, 'You must stick to your plan. You must carry out orders.' On the left of the advance – around Gommecourt and Beaumont Hamel – the British made no progress at all and suffered dreadful casualties. In the centre, behind which Gough's Reserve Army of three cavalry and two infantry divisions was massed, some progress was made, but successes were isolated. Even where German trenches were taken the forward troops could not communicate with the British lines to tell them of their success as runners were shot down in hundreds and telephone wires cut by artillery fire. Heroism was present all along the line, and was notably displayed by the 36th Ulster Division which took the Schwaben Redoubt in front of Thiepval and held it without reinforcements until driven out by German counter-attacks, having suffered over 5,000 casualties. Only on the right, where better French tactics prevailed, was real progress made. The Germans were taken by surprise here, for they had not expected the French to be able to launch an attack in view of their recent martyrdom in the great struggle around Verdun.

By the end of the day the British held a three-mile-wide portion of the German position, to a depth of one mile, and just three of the 13 target villages. At no point had they reached the second line of German defences. Haig's breakthrough was out of the question – the cavalry would not be needed. Rawlinson had more reason to be pleased than Haig – his men were at least 'holding' a small sector – but it had been achieved at the cost of nearly eight casualties for every one German. Haig blithely commented that the casualties could not be considered severe in view of the numbers engaged. Yet 50 per cent casualties were so rare in military history that they usually represented a defeat as decisive as the French had suffered at Waterloo. No previous army commander with such casualties had ever expressed himself satisfied. Haig's complete ignorance of events was clearly demonstrated by his unforgivable attribution of cowardice to Hunter-Weston's Eighth Corps. In Haig's words, 'few of the VIII Corps even left their trenches'. In fact, Eighth Corps suffered over 13,000 casualties – the highest of any corps involved. And what was awaiting the casualties who were lucky enough to be brought back from the front? Between Albert and Amiens a casualty clearing station had been set up to expect 1,000 casualties. Within a few hours of the start of the battle they were overwhelmed by 10,000 wounded men. A surgeon there wrote of 'streams of ambulances a mile long' waiting to be unloaded.

The five months of fighting on the Somme were to see the most bitter attritional battle not just of the First World War but in all history. The British and the Germans were to fire over thirty million shells at each other and suffer in return over a million casualties in an area little more than seven square miles in extent. It was the biggest abbatoir ever devised by man.

'The French! They're the fellows we shall be fighting next.'

Field Marshal Haig in 1919

Four and a half years of fighting in alliance with the French seems only to have confirmed Haig's view that they rather than the Germans were Britain's real enemy. The insularity and chauvinism of many British generals, notably Haig's predecessor as commander-in-chief, Sir John French, prevented Britain from ever being really 'willing' allies.

THE BATTLE OF CAPORETTO (1917)

Italy entered the First World War in 1915 on a wave of public enthusiasm but without any serious military preparation. The fact that Italy had been negotiating simultaneously with both the Central Powers and the Entente made it impossible for the Italians to establish military priorities or work out plans of campaign. Italy's commander-in-chief, Field Marshal Luigi Cadorna, was not a gifted soldier and owed his prominence more to the achievements of his father – a hero of the fight for Italian unity – than to any accomplishments of his own. Other Italian generals were of poor quality – certainly the worst of any of the combatant nations. Ex-Prime Minister Giolitti called Italy's commanders worthless, on the grounds that most had joined the army when it was the fashion among rich Italian families to send their least gifted sons into the armed forces.

Cadorna's own personality was a large part of the problem. He was not a man to admit his mistakes and was as arrogant as most of the Italian officers who belonged to what was known as the 'Piedmont clique'. Although at the inter-Allied meetings – when he met Joffre and Haig – he had always seemed happy to agree to coordinate his attacks with those of his western and eastern allies, it was notable that his attacks did not begin until those of his allies had already stopped. Cadorna was basically not in sympathy with trench warfare and thought in terms of heroic, but archaic methods of fighting. He gave his generals *carte blanche* to attack when and what they liked, provided that it took Italian troops 'forward'.

Cadorna's 'egocentric authoritarianism' was at the heart of Italian incompetence. His style of leadership led to poor intelligence, staffwork and coordination, and he was usually far too distant from the scene of the fighting. He allowed slackness among his generals, taking no action against senior commanders who still tried to run a war as a part-time affair while pursuing more relaxing hobbies. The 3rd Army seemed to be run on a part-time basis by the Duke of Aosta, who delegated responsibility to relatively junior men. One of his corps commanders, General Pecori Giraldi, never once visited HQ during the winter of 1916-17 but stayed at his villa and sent cars to fetch his staff for meals. It was a cosy arrangement, but a far cry from the Spartan lifestyles of some of the German generals.

While Cadorna had his favourites, who could do no wrong, those not directly within his circle were shuffled around and sacked with impunity. By October 1917, 217 generals, 255 colonels and 355 battalion commanders had been sacked – not all for incompetence; many, in fact, for being too keen or too able. The highly inept General Carignani once warned his senior brigadier – an able man – 'I know I have enemies; I know I may fall; and if I do, so will a lot of other people. So remember that.'

The Italians had received ample warning that the Austrians were preparing a big attack, but Cadorna simply dismissed all the evidence on the grounds that the enemy would not attack just before the winter set in, or in the mountain zone. He would not do so, and therefore they would not do it either. That was all there was to say on the subject. But his confidence was ill-founded, for that was precisely what the Austro-Germans were planning to do. The Italians had been asking for trouble when their prime minister, the decrepit 75-year-old Paolo Boselli, reached a fateful decision. Italy had declared war only on Austria in 1915 and in the following two years had not officially been at war with Germany. However, on 17 August, 1917, Boselli 'bit the bullet' and declared war on Germany and Turkey, hoping to be able to gain territory from them when the war was over. It was a decision Italy was to regret. The Germans now briefly turned their attention to Italy and decided to help their Austrian allies to knock her out of the war with one tremendous blow. Field Marshal Erich von Ludendorff was sent to oversee a great offensive in the Dolomites, which would be spearheaded by fifteen infantry divisions, including the eight élite German divisions of General von Below's 14th Army. The Italians were in for a shock. The cosy – if bloody – war against their fellow incompetents the Austrians was about to be taken by the scruff of its neck and shaken. Operation *Waffentreue* ('Brothers-in-Arms') was about to burst on them.

In the early hours of 24 October 1917, the battle of Caporetto began with a volcanic bombardment from Below's 14th Army, mixing high explosives with gas shells. The Austrians knew well that the Italian gas masks were quite inadequate and realized just how effective gas would prove on this front. They could advance, confident in the knowledge that large numbers of Italian troops would be incapacitated. Typically there was no Italian counter-fire – one officer refusing to use his guns in order to conserve ammunition – and the Austrians knew from long experience that there would be none. They also believed that once the Italians suffered a setback they would collapse like a house of cards. They were right. Cadorna, controlling the battle from Udine, 20 miles from the fighting, decided he would be safer further back and retreated a hundred miles to Padua. It was not impressive leadership and it set the pattern for generals, officers, soldiers and civilians to run away.

To add to their woes the Italians had absolutely no luck. The 2nd Army commander, General Luigi Capello, probably Italy's best fighting soldier, had been away from the front undergoing specialist medical treatment. No sooner had he returned than the attack began and he collapsed, suffering from fever. Admitted to hospital he insisted on retaining control of 2nd Army from his sick bed – with farcical results. As the Austrians drove back the Italian 2nd Army, morale collapsed among the Italian troops, who shouted 'blackleg' at reserves moving to the front. Austrian advanced units met Italians shouting, '*Eviva La Austria.*' Demoralization had set in.

As the Austro-German forces pressed forward, the young Oberleutnant Rommel, alone and on foot, encountered a large group of Italians and told them to surrender. On seeing him walking towards them, they pushed their officers aside and ran to congratulate him, lifting him on their shoulders and saying, '*Eviva Germania.*' An Italian officer, less willing to surrender to a solitary German, was shot down by his own troops. Sometimes Italy seemed to be at war with herself. Single-handedly (actually with two officers and three riflemen) Rommel captured 1,500 men and 43 officers from the 1st Salerno Brigade.

A problem that the Austro-Germans did encounter was their inability to gather weather information as they could not launch their aircraft in the dense, low clouds. But the Italians kindly supplied all the information they needed through their weather forecasts on the radio.

The 2nd Army had simply disintegrated, with most of its men heading home to get into civilian clothes as quickly as possible. Whole Italian divisions disappeared from the front, along with their generals and staff officers, while their artillery was only notable for being in the forefront of any scramble to escape. But as the Italian army fled south, vast numbers of refugees – as many as 500,000 – fled with them, blocking all the roads with their horses, carts and other paraphernalia. In fact it was these refugees who turned out to be Italy's best soldiers, holding up the Austro-German advance more effectively than Italy's army had done. As they fled, the refugees descended on peasant villages like a swarm of locusts, looting and taking all the food and livestock. The Austrian advanced troops found not just the roads but the fields alongside packed with millions of people, soldiers and civilians, so that it was impossible to take any more prisoners.

The Italian reverse at Caporetto was an astonishing event; more than a military defeat, it took on all the appearances of a major natural disaster, affecting a whole population. It was the most stunning victory of the entire First World War. The elimination of the 2nd Army, combined with terrible losses elsewhere, meant that the Italian army suffered 800,000 casualties, most from desertion. For a loss of just 5,000 men the Austrians and Germans had scored what should have been a war-winning victory. What saved Italy was the inability of the enemy to exploit it. Ludendorff and von Below had instigated *blitzkrieg* but they lacked the Stukas and the fast-moving tanks of the Second World War to make their breakthrough decisive. Eventually, with a stiffening of British and French divisions from the

'As to the battle of Arras, I know quite well that I am being used as a tool in the hands of the Divine power and that my strength is not my own.'

General Haig in 1917

Unconvincing evidence of a benevolent deity.

Western Front, the Italians established new positions on the River Piave and the panic subsided. But for a while it seemed as if an avalanche had streamed from the mountains to crush Italy.

THE BATTLE OF PASSCHENDAELE (1917)

Victories are usually judged by territory gained, targets achieved or enemy soldiers killed or captured. By any of these standards the battle of Passchendaele (otherwise known as 3rd Ypres) was a curious kind of victory. But according to General Douglas Haig a victory it was. Britain had gained an area of Belgian mud that would keep his stout walking boots busy for several hours. And nearly a quarter of a million Germans – give or take a few tens of thousands – had been written off in the process. That was progress enough. Admittedly the targets had been a touch optimistic, but two out of three was enough to make Passchendaele a victory. And all the luck had been going Germany's way. Who could have predicted such wet weather? And by what chance had the Germans put all their strongest defences at the very place where the British wanted to attack them?

But it was not a matter of luck. The British had known for some time that the German defences at Passchendaele were very strong. Intelligence reports since autumn 1916 had reported unusual German activity in the area and half a million aerial photographs of the area showed every possible detail of the German defences. But these revelations presented Haig with a dilemma. In order to break down the German defences he would need a heavy preliminary bombardment, even greater than that used on the Somme. However, the area was close to sea level and would flood very easily. Once the big guns had churned up the ground the area would be impassable for tanks, leaving the infantry to slog through the slime on their own. If the weather stayed dry, Haig reasoned, it just might be done. But if it rained, the whole front would soon become impossible. He therefore had the weather charts for the region checked and rechecked. Over the last 30 years July and August had been unpredictable months with occasional heavy thunderstorms. But September was usually good, so if you were lucky and had a fine July and August you might have a stretch of twelve dry weeks. But that was as much as you could expect, for October was the wettest month of the year. Any campaign started in July must not go on beyond September. In this way Haig was fully informed about the German defences and about the likely weather he could expect. And he knew that once started the battle must be concluded quickly; there could be no long drawn-out attrition as on the Somme in 1916. Appreciating the need for quick action Haig chose the 'dashing' Hubert Gough to spearhead the assault rather than the more reliable, but ponderous Plumer.

By 1917 the Germans had mastered the art of 'spoiling' Allied attacks by the use of artillery interdiction and the new mustard gas, for which the British had no antidote and little protection. Against Nivelle's offensive only months before the Germans had responded with a simpler and even more frustrating tactic. Before the Allies could open their heavy preliminary bombardment the Germans simply pulled back their troops, leaving the enemy to bombard a void. Having devastated the ground, British and French troops had to advance through it and presented easy targets for German artillery. It was cunning and very effective. Haig, in frustration, tried to pin the blame for the Germans' knowing when the British attack would begin on a wretched Welsh sergeant, captured by the Germans, who broke down under interrogation and 'blabbed'. Yet if the Germans had to ask a captured soldier their intelligence forces must have been asleep, for Field Marshal Robertson wrote the night before the attack that, 'everybody in my hotel knows the date of the offensive down to the lift boy'.

Frankly Haig was out of his depth. As the war went on he dug in his heels, doing the same thing over and over again in the hope that it would eventually work. Haig placed his faith in British 'stickability'. Let the Germans and the French tinker with tactics, Haig believed that the British soldier – with his back to the wall – would see it through. But at what price?

The British assault began on 31 July and as a preliminary to the attack British guns had turned the terrain into a quagmire by depositing five tons of shells on each square yard of Flanders mud. The Germans had been impressed at this display of pyrotechnics, thanking their lucky stars that none of

British troops crossing the Ypres canal near Boesinghe, 5 August 1917, during the early stages of the battle of Passchendaele, a battle whose name has become synonymous with the horrors of trench warfare.

it had come anywhere near them. The subsequent British infantry attack was more like an amphibious operation, except the soldiers had no boats. As they sank into the swamp they provided easy targets for the German machine guns, firing from massive concrete bunkers standing out like martello towers in the sea. Four days after the attack began rain was stair-rodding down. Haig was beginning to doubt the wisdom of going on. His only chance had been a quick breakthrough. If only he could have accepted that he was wrong and called off the operation before too much damage was done. All that he was left with was the chance of a 'slogging-match' like the Somme, but worse. Were the British in a condition to fight an attritional battle so soon after the Somme? The answer should have been a resounding 'no'.

First there was the problem of numbers. By the end of 1917 – with Russia out of the war – Germany could expect to transfer as many as 20 new divisions to the Western Front. Intelligence reports also indicated that the Germans had a million men in training depots and a further two million available should they require them. All this had to be set against a British shortfall of 60,000 men in France, and Lord Derby's startling revelation that British manpower was already 'at bedrock'. Derby made it absolutely clear to Haig that manpower must be conserved at all costs. This was hardly the support that Haig needed if he was planning another attritional battle. Even British divisions in the line in the summer of 1917 were often up to 4,000 men below strength. Haig was already cooking the books by clearing V.D. hospitals and ordering lightly wounded men back to the front. In London ministers were becoming very suspicious of Haig's low casualty returns. Where were the lightly wounded men going?

If the manpower situation was so serious why did Haig persevere with the Passchendaele offensive after it had demonstrably failed to achieve a quick breakthrough? Clearly he did not have the human resources for a long battle and so why did Lloyd George not stop him – even sack him? The truth lies in the realm of grand strategy. Britain was aware that Germany had made peace offers to France, but the British were unwilling to see their ally accept the offers before they had had one last attempt to drive the Germans away from the North Sea coast of Belgium. Haig, on the other hand, was neither so

Machiavellian nor so coherent in his thinking. His reports to the cabinet during the battle of Passchendaele contain embarrassing blather about the superiority of British artillery and the low German morale. This convinced nobody. But Lloyd George, far from holding back soldiers and trying to prevent Haig squandering lives, was apparently prepared for Haig to have one last go at shifting the Germans from the coastal region of Belgium.

And so the greatest blunder of the Passchendaele fighting – which alone cost 100,000 casualties – took place in October and November, not just because Haig did not know when he was beaten, but because the British government was willing to squander lives in pursuit of a victory, in case a negotiated peace was arranged between France and Germany. Haig was a suitable scapegoat should the plan go wrong – as they acknowledged that it probably would.

In order to continue the fight into October and November 1917 and try to take Passchendaele village itself Haig decided to use his best strike troops – Australians and New Zealanders. Their target was to take the village by 12 October, after which the Canadians and the Cavalry Corps would push on to the railway junction at Roulers and cut the Germans off from the coast. It sounded easy, but the condition of the ground made it simply impossible.

Two British divisions – 49th and 66th – were earmarked to carry out preliminary work for the Australians. Unfortunately, on 10 October catastrophe struck the 66th. It was an inexperienced division which had not been properly briefed about its task. It arrived late, having spent nearly 24 hours on the march, without food or water. The covering barrage for the 66th was far too far forward and had to be ordered back. But the gunners were out of touch with the front line and as they brought back their barrage it struck the unlucky 66th, cutting the division to pieces and inflicting heavy casualties. The Australians had witnessed what had happened and their commander, Birdwood, complained to GHQ that he had just seen British soldiers wiped out by their own gunners. But the matter was hushed up. The British were less willing to acknowledge the existence of friendly fire than their French allies.

Haig was confident that the Australians could achieve their mission, which was to take Passchendaele village. 'The New Zealand and Australian 3rd Division are to put the Australian flag in the church there,' he told his wife. But there were no flags and no celebrations; the attack was a disaster, with the Australians suffering 60 per cent casualties. Even Haig was prepared to admit that the ground now was 'quite impossible'. But the battle went on. Next it was the turn of the Canadians. But over half of their guns were underwater or clogged with mud or just 'missing', presumed drowned. Lieutenant-Colonel Alanbrooke – later to distinguish himself in the Second World War – attended one of Haig's briefings for the Canadians and could hardly believe his ears. Haig spoke as if the attack was taking place in high summer up a wide, dusty road with the enemy in craven retreat. It was lies, damned lies, and false statistics. Alanbrooke suggested that Haig either had never seen what it was like at the front, or else his capacity to keep a straight face while he lied was unsurpassed.

Denis Winter has described the reaction of Canadian Prime Minister, Sir Robert Borden, to what the Canadians had to endure. It so infuriated him that at one meeting of the Imperial War Cabinet he took Lloyd George by the lapels and shook him in a fury. Apparently, as a result of inaccurate orders two companies of the Canadians had been positioned 100 yards ahead of the artillery barrage and as it was advanced both were blown to pieces by the British guns. Then the Canadian survivors, running back from the slaughter ahead, were attacked in error by two British companies supposedly supporting them. When the Canadians were informed that their assault was to be backed by 364 guns, they decided to carry out a check and found that there were in fact just 200 guns to support them. When the Canadian commander complained and demanded more guns he was told that he would have to send an indent to GHQ as there were no more available.

Not that life was any easier for the much-maligned gunners. With the rain incessant through October, one gunner reported that in his battery the water was as deep as the gun's breech, which went underwater every time the gun was fired. The gunners tried to convince GHQ that if the battle went on there would be no heavy guns at all for use in 1918. Colonel Rawlins, artillery adviser to Haig, actually had the temerity to make this view known to the great man. Haig went white and shouted, 'Colonel Rawlins, leave the room!' When Brigadier-General Edmonds tried to support Rawlins,

Haig turned on him, 'You go too, Edmonds!'

The capture of the village of Passchendaele by the Canadians provided a suitable way of ending the battle. It had been a victory. Ground had been gained and many casualties inflicted on the enemy. Admittedly British casualties had been high – in truth closer to 350,000 than to the 238,000 given by the Official History. But what of the enemy? At the outset Haig had known that he had no men to spare, but that had not stopped him sacrificing another third of a million of them even when he had heard that Britain's manpower was at bedrock. What had it all been for? Politically, of course, there had been targets unknown to the fighting man. But the Belgian coast had not been taken by the British, nor the German flank turned. And Germany was not driven by her losses to seek a peace on terms favourable to Britain. In fact, the Germans had lost considerably fewer men than the British – as few as 200,000 perhaps – in an operation that had been intended to sap their will to continue. All Haig had done was to demonstrate his – and Britain's – incapacity to win the war. Britain would now have to 'sit tight and wait for the Americans' to redress the balance on the Western Front. The battle of Passchendaele, for all its secret political agenda, had been what everyone believed all along, a doomed offensive that should have been aborted as soon as its faint trickle of life began to fade, just days into the offensive.

'Battles cannot be stopped like tennis matches for showers.'

General Charteris, replying to Lloyd George's comment on the appalling fighting conditions at Passchendaele in 1917

A callous comment from Haig's head of intelligence. Charteris was the *eminence grise* behind many of Haig's most ill-considered and ill-informed utterances.

THE BATTLE OF ANUAL (1921)

Birthdays can be difficult times, particularly kings' birthdays. It is always so hard to know what to get the man who has everything. Yet during the summer of 1921, Spanish General Manuel Silvestre thought he had the answer to this pressing problem. He would present King Alphonso XIII of Spain with a large part of Morocco, known as the Rif. But when the Spanish High Commissioner, Damasco Berenguer, refused to sanction the gift on the very reasonable grounds that the territory belonged to the Rifs, Silvestre went berserk and tried to throttle him. Silvestre was inclined to be temperamental. As a 'fighting general', wounded 16 times in the war against America in 1898, he was a man whose courage was beyond question even if his judgment on military affairs certainly was not. He was a favourite of the Spanish court, a close personal friend of King Alphonso, a raconteur, a master of the social graces, a connoisseur of fine wines – in fact, all that could be expected of a Spanish general. But was he suited to commanding a second-rate colonial army in Morocco?

During their colonial occupation of Morocco in the period after World War I, the Spanish had adopted a policy of sprinkling small garrisons of 20 or so men throughout the country, placing them in hundreds of blockhouses and forts. The result of such isolation was that the army suffered from low morale, with ordinary soldiers being poorly fed and enduring living conditions which would have been considered a scandal in Europe. The Spanish soldier's pay was very low – Berber tribesmen earned three times as much for labouring on the roads. Most of the conscripted troops were illiterate agricultural workers who had not been trained to handle modern equipment. When given a modern rifle, some immediately sold it to local Rif tribesmen so that they could buy better food, cigarettes or wine. In one garrison, 19 of the 30 soldiers had

Goering and Hitler look on as French prime minister Edouard Daladier signs Hitler's guest book during the Munich conference, 29 September 1938. Daladier and Chamberlain, the British prime minister, believed they had won 'peace in our time', but they had only delayed the inevitable, giving Germany a further twelve months to re-arm.

even more concerned that Britain should avoid bankruptcy, and in his view re-armament in the mid-1930s would have spelled financial ruin. Yet it was not only financial reasons that prevented Britain re-arming. Britain was also fearful of her vulnerability to aerial attack and seemed willing to do almost anything to avoid war. The result was that from 1934, instead of re-arming Britain adopted a strategic policy of 'deterrence'.

The British decision to adopt a policy of deterrence was based on the theories of Trenchard in Britain and Douhet in Italy that in the next war strategic bombing of urban areas would prove irresistible and that active defence against the bomber was impossible. The conclusion reached by military leaders in Britain was that the only defence against such a terror weapon was the capacity to strike back just as hard. Thus Britain would purchase security at the cost of equipping her air force with heavy bombers – a low cost compared with maintaining a large standing army. Chamberlain expressed British policy in these words:

> . . . our best defence would be the existence of a deterrent force so powerful as to render success in attack too doubtful to be worthwhile. I submit that it is most likely to be attained by the establishment of an Air Force based in this country of a size and efficiency calculated to inspire respect in the mind of a possible enemy.

The British Chiefs of Staff did not agree. They asserted that only a large army, capable of intervening on the Continent, would deter an aggressor. To them, both potential allies and potential enemies would view the British concentration on an air force as typical of Britain's desire to avoid a European commitment. Thus the value of the deterrent would be low. They pressed for the creation of a British field force as well as an expansion of the navy to reach a two-power standard to match both Germany and Japan. However, their plans fell victim to Treasury logic.

The re-militarization of the Rhineland by Hitler in 1936 came as no surprise to the French, who believed war with Germany could not long be delayed. They tried to persuade Britain to give priority to the creation of a sizeable expeditionary force to fight alongside the French army. But they were doomed to disappointment. Britain felt secure with her deterrence policy. Admittedly France, with a vulnerable land frontier with Germany, might succumb to a resurgent Germany; but while Britain could threaten to destroy Germany's cities, Chamberlain and other Conservative politicians felt safe. But they were mistaken. Britain's policy of deterrence fooled nobody. German intelligence sources were well aware that Britain did not have the bomber force necessary to inflict really decisive damage on Germany's war effort. In 1938, no country, not even Germany, had the capacity for long-range civilian bombing. Deterrence failed to deter Hitler because it was not backed up by military reality. Britain had been unwilling to accept the economic consequences of financing re-armament, which is why she had originally opted for the less costly policy of deterrence through strategic bombing. Yet how could her enemies regard her as a realistic obstacle to their ambitions when she was also unwilling or unable to back up her policy with the bomber force that would deter them? The only policy that would have deterred Hitler was if Britain had committed herself to building up a large army, ready to fight alongside the French. This would have deterred Hitler from his adventures in Czechoslovakia and Poland because he could not have risked a decisive Allied strike on his undefended western frontier. As it was, without any hope of substantial British military aid, France became committed to the defensive and Hitler correctly read the signs that the French army had lost its will to fight. The Maginot Line thus served more as a prison than a fortress for French hopes.

In 1938 Hitler's claims to the Sudetenland brought Europe to the brink of war. It seemed for a while that Britain and France would support the Czechs by force, but at the Munich Conference a deal was made that postponed war for a year. Historians have been divided as to the wisdom of the continued appeasement of the German dictator. Some argue that Germany was at her weakest in 1938 relative to Britain and France and that war in that year could have been a strategic disaster for Hitler. Other historians, however, point out the pitiful state of Britain's air defences in 1938 and feel that she would have succumbed to the offensive power of the Luftwaffe.

It is common misconception that the Luftwaffe could have inflicted heavy damage on British and French cities in 1938. In fact it was in no position to launch or sustain such an attack and was being developed as a tactical air force, designed to support fast-moving tank and motorized infantry units. Its concentration on dive-bombers and medium-range bombers showed that it was not German policy in 1938 to undertake the strategic bombing campaign that everyone feared. Ironically, the Germans had not succumbed to the myth of the bomber in the way that the British had.

German military strength in 1938 was a shadow of what it would be in eighteen months' time. At the time of the Munich crisis the German army had 48 regular divisions, of which only three were armoured, four were light reconnaissance and four were motorized infantry. It lacked heavy artillery and had few reserves. It should also be noted that five of the divisions were Austrian and of a far lower standard than the German units. The three armoured divisions were equipped with the obsolete PKW I and II light tanks, while the medium tanks PKW III and IV were available only in small numbers for combat testing. The bulk of the German army consisted of 37 infantry divisions, little changed from 1918, with horse-drawn artillery and transport. In the event of a crisis Germany could be expected to mobilize eight reserve divisions and 21 Landwehr divisions, mostly made up of aging First World War veterans who possessed little recent training or modern equipment.

Naturally it is wrong to over-estimate the strength of the Czech army, which was a 'paper' army only and had not been tested in action. Of the 30 Czech infantry divisions only nineteen of these belonged to the regular army and the reserve divisions lacked modern equipment. Yet the regular divisions may be considered as equal to their German equivalents and were equipped with excellent weapons. The Czechs had built extensive fortifications on their Silesian frontier and might have held up the Germans for some time, although with the poor quality of their military leadership it is doubtful if anything further would have been achieved. Nevertheless, the Germans would have had to fight, rather than occupying a country surrendered to them without a struggle. Casualties would probably have been high, progress slow and victory hard earned. And most important of all, the Germans would not have been able to capture Czech equipment intact and use it themselves in 1939. Had the Germans been forced to fight their way into Czechoslovakia it is unlikely that they would have been able to take over a viable armaments industry as they did.

It has become the fashion to dismiss the potential of the French army, possibly because Hitler did so and was proved to have been right. However, the German generals did not place much trust in Hitler's intuition and preferred to base their strategy on firmer foundations, like intelligence reports. It was difficult for them to believe that the nation which had fought with such determination in the First World War would fold up so easily in 1940.

The myth of German armoured superiority in May 1940 has been effectively disproved. There can be no doubt that in 1938 the Germans would have found the French mechanized forces formidable opponents. The French tank industry had begun mass production in 1936 and produced more tanks of greater size and sophistication than the Germans. The French were even outproducing German factories by 150 tanks to 100 each month. Properly used, the French tanks would have achieved at least a stalemate in 1938 and probably far more. Only the acquisition of Czech tanks and production facilities enabled the Germans to create the four new panzer divisions which employed 297 Czech TNHP 35 and 38 tanks to replace the obsolete PKW Is.

The situation in the Soviet Union is also relevant. In September 1939, when Hitler invaded Poland, Stalin had completed his purge of the Red Army and was an ally of Germany. A year earlier, however, he was still in the process of eliminating dissidents from the army and therefore the military strength of Russia was small. Nevertheless, in 1938 he had not yet allied to Germany and was not supporting her economy with large quantities of raw materials as he was after the Nazi–Soviet Pact. Even if Russia was no military threat to Germany, Hitler could not count on getting any support from her. In addition, Romania was alarmed by the threat to Czechoslovakia and warned Hitler that if he continued his aggression he could expect no more Romanian oil after October 1938.

Whatever plans Hitler had for Eastern Europe, he could not overlook the fact that his main enemies were in the west. Germany's 'Maginot Line' – the Westwall – was a 'complete sham', which had only been started in 1938 and by the time of Munich had just 517 bunkers, most of which were militarily useless as their concrete had not yet set. Significantly, a year later the figure stood at 10,000. In the absence of formal defensive positions it would have been necessary to employ large ground forces on the French frontier and in view of the potential strength of the Czech army where would Hitler have found these? In 1938 General Adam commanded the entire western front with just five regular divisions and although Hitler promised him a further 20 if war broke out with France and Britain, Brauchitsch immediately contradicted the Führer and said that only eight could be made available in the first three weeks.

Hitler was a gambler and did not allow such problems to worry him. He believed the French army was weak, for all its apparent size, and that its will to fight was negligible. He was right, but it was intuition rather than intelligence reports that convinced him. He found the French military leaders passive and fearful. The French commander-in-chief General Gamelin had no intention of launching an attack on the Rhineland if Germany attacked Czechoslovakia. He was prepared only to occupy the Maginot Line. In spite of a numerical advantage in the west of 56 divisions against an estimated eight German divisions, Gamelin apparently lacked confidence in France's ability to win a worthwhile victory.

In Britain Chamberlain had nightmares of the horrors of German bombing. The Imperial Defence Committee had told him that the Luftwaffe would drop 3,500 tons of bombs on London in the first 24 hours, while the Ministry of Health told him to anticipate 600,000 killed and 1,200,000 wounded in the first six months. These estimates shocked Chamberlain, but they were groundless. During the whole of the war there were only 60,000 civilian deaths from bombing. Fear was at the root of appeasement over Czechoslovakia.

The truth was that in 1938 Hitler lacked the bombers to destroy civilian morale because he had concentrated his resources on winning the war on the ground, building panzer tanks and Stuka dive-bombers. With hindsight it is clear how little the Allies had to gain by delaying the inevitable. Chamberlain may have believed he had achieved 'peace in our time', but his military advisers should have convinced him that he was flying in the face of reality. As Churchill said, 'The year's breathing space said to be "gained" by Munich left Britain and France in a much worse position compared to Hitler's Germany than they had been at the Munich crisis.'

In 1938 the balance of naval power, traditionally vital to Britain, was remarkably favourable. Neither of the German battlecruisers, *Scharnhorst* and *Gneisenau*, had been completed, nor had the great battleships *Bismarck* and *Tirpitz*. Germany possessed no heavy cruisers, no aircraft carriers, just six light cruisers, seven destroyers and only twelve ocean-going U-boats. The German Naval High Command (OKM) doubted its capacity to even protect the trade routes in the Baltic and the Swedish iron ore shipments. Even without the King George V class battleships, Britain possessed a far greater mastery over Germany than she was to enjoy twelve or eighteen months later.

In 1938 the German economy was extremely vulnerable to disruption. Supplies of vital iron ore from Scandinavia could be cut off as they had been in the First World War by British naval blockade. And although Germany was self-sufficient in coal, her western coalfields, particularly those in the Saar, were vulnerable to French military action. Efforts to create a synthetic fuel industry in the 1930s had not succeeded in making Germany self-sufficient in oil products. Far from it, for in June 1938 Germany had only enough petrol stockpiled to cover 25 per cent of mobilization requirements, while the situation for aviation lubricants was even more desperate: just 6 per cent of wartime requirements. The problem with rubber was just as bad for in mid-1938 synthetic rubber production provided just 6 per cent of German needs. In terms of munitions, gun powder production was just 40 per cent and explosives 30 per cent of the First World War maximum figure. Hitler faced a stark threat of economic defeat in 1938, and in the period between Munich and the outbreak of war in September 1939 Germany made prodigious efforts to make up the shortfall.

The truth is that none of the European powers was prepared for an all-out war in 1938. Hitler knew that if he went to war with Czechoslovakia he might bring down on himself not just the strength of the French army but blockade by Britain's navy which could spell economic ruin. Yet he was a gambler by nature and believed that he could exploit the fears that had come to dominate his enemies: civilian bombing in the case of Britain and invasion for the third time in 70 years in the case of the French.

At the Munich Conference Chamberlain and Daladier preferred to stab an ally in the back rather than face an enemy. Their treachery earned them a twelve-month reprieve. Yet time was no friend to the Allies. Germany made better use of the twelve months' delay and was a far more formidable enemy in 1939 than she would have been if Britain and France had supported the Czechs at the time of the

Field-Marshal Erwin Rommel commanding the German Afrika Korps during the Second World War. Notwithstanding the inexperience of American troops at the Kasserine Pass – and the incompetence of their generals – the threat of a British advance from the south persuaded Rommel to pull back from western Tunisia.

none of the flair of the 7th cavalry and 'Gentleman Jim' Alger was no Errol Flynn, but he would give it a go. Alger did not know it, but he was the main course at the barbecue. With a hotchpotch of tanks and guns, no air cover and no reconnaissance, he was being sent straight out to fight 100 German tanks, supported by dive-bombers and heavy artillery. In all he lost 50 out of his 54 tanks and was himself captured. The movie was over – there was to be no eleventh-hour rescue.

Fredendall's conduct of affairs had been quite unbelievable. How Alger – with just 54 medium tanks – could be expected to rescue infantry trapped over thirteen miles away by taking on two panzer divisions, while overhead Stukas dived and bombed him to oblivion, is hard to imagine. Fredendall had no option now but to drop leaflets on the trapped GIs on the hills telling them to try to break out and make their own way back. That evening Colonel Drake led his 1,600 men down the slopes in the darkness and tried to get away but they were quickly surrounded by German tanks and forced to surrender. To the north Major Robert Moore was luckier: 300 of his 900 men made it back to American lines, but the rest were captured or killed in the attempt. Fredendall did not like the script. In two days he had lost two battalions of armour, two of infantry and two of artillery, and at scarcely any cost to the Germans. Whose side were these film directors on?

Eisenhower had no choice now but to order II Corps back 50 miles to defend the Kasserine Pass, gateway to the strategic Algerian town of Tebessa. Should Rommel capture Tebessa he would have a free road north to the Algerian coast, disrupting the rear of the whole Allied position in North Africa. But the leadership of II Corps was disintegrating. Morale had slumped and panic was never far from the surface. Fredendall himself was at odds with his own commanders. Yet by chance the Axis commanders von Arnim and Rommel were also at daggers drawn. Their disagreement over strategy gave the Americans an unexpected two days' breathing space. Rommel appealed to Field Marshal Kesselring in Rome to allow him to press his attack against Fredendall, with the aim of taking Tebessa. But von Arnim wanted to keep much of the German heavy armour for an operation of his own. Eventually Kesselring gave Rommel the go-ahead. The delay had given Fredendall the chance to reform his shattered command, but yet again he had parcelled out his troops so thinly that they would be unable to hold any of the five passes through which Rommel might choose to advance.

In the Kasserine Pass Fredendall had instructed the 26th Infantry Regiment commander, Colonel Alexander Stark, to 'pull a Stonewall Jackson' (Confederate General Thomas Jackson had won the nickname 'Stonewall' for his heroic defensive performance at the First Battle of Bull Run in 1861). This was excellent advice in principle, but what kind of troops did Stark have to hold the pass? The answer was green GIs and a bunch of engineers who had never been under fire before. It was asking a lot of men handier with a spanner than a sten gun to stand against the battle-hardened German troops. And it was easy for Fredendall to give instructions to 'pull a Stonewall' when he was 70 miles from the action and 20 feet underground at that.

On 19 February Brigadier-General Buelowius's panzers tried to 'bulldoze' their way through Stark's defences in the Kasserine Pass, but the Americans gamely held them off with artillery, anti-tank and small arms fire. It was not until the veteran Afrika Korps infantry got round Stark's flanks and attacked him from the rear that the rot set in. Stark's force lost their nerve and cracked, and another humiliating retreat began. On the road to Tebessa a company of engineers panicked at what they assumed, in the shadows, to be approaching German tanks. The panic spread and soon an atmosphere of hysteria prevailed among the American troops. Forward artillery observers fled, declaring, 'This place is too hot.' As darkness fell hundreds of men left their posts and began to run. The Kasserine defences were caving in. Reinforcements of American infantry and British tanks were rushed in to hold the Germans for a while but by the following afternoon Buelowius had broken through. It seemed now that nothing could stop Rommel from breaking out into open country.

But Rommel, suffering from jaundice and desert sores, was a prey to doubt. While he had not been impressed by the American soldiers he had encountered so far, he knew that they would get better with training and experience. And what broke his spirit – and plunged him into despair – was the seemingly limitless supplies the Allies had. How could he maintain the fight against an enemy who could replace everything he had lost – men, guns and tanks? And at any moment he knew that Montgomery's 8th Army would break through into southern Tunisia. Could he afford to have his old enemy operating in his rear? He had no choice and decided to pull back from western Tunisia.

The Americans were to have the breathing space they needed to improve. And Eisenhower made a start by replacing Fredendall with George Patton. Patton needed no concrete bunker in a secluded ravine; a tank would do, out front of the rest of them. And if Patton did not like the script he would write the whole damned film himself.

THE BATTLE OF VILLERS-BOCAGE (1944)

It has become fashionable to undervalue the role of the individual in military history. Surely no individual, however brilliant, can affect the outcome of a battle by his own particular efforts. Yet on 13 June 1944 Captain Michael Wittman, of the 501st SS Heavy Tank battalion, 'won' the battle of Villers-

British troops fighting in the hedgerows outside Caen, July 1944. Many British soldiers, accustomed to desert fighting, found conditions in northern France extremely difficult. British tank crews were outperformed by ace German units with experience of fighting on the eastern front.

Bocage virtually single-handed, and blunted Field Marshal Montgomery's push towards Caen. It is difficult to recall another occasion in military history where one man so influenced the result of a battle – or, to put it another way, it is difficult to remember such supine resistance by a column of tanks which allowed one man to destroy it.

The battle of Villers-Bocage represents a low point for the British Army in the Normandy campaign, as well as a particularly sad period for the 7th Armoured Division – part of Montgomery's immortal 'Desert Rats'. The division's own historian agrees that 'the normally very high morale of the Division fell temporarily to a very low ebb . . . A kind of claustrophobia affected the troops . . .'

Field Marshal Montgomery's plan to take Caen had originally allowed for an airborne drop by 1st Airborne Division behind the town as soon as 7th Armoured and 51st Highlanders had pushed up from the coast. However, the complete failure of 7th Armoured to make that push put an end to such planning. Major-General Erskine, commanding the 7th Armoured Division, had initially been excessively optimistic, reporting limited German resistance and the loss of just four tanks. Yet the division was moving far too slowly and was failing to give armoured support to the infantry. It was on 11 June that XXX Corps commander, General Bucknall, detected a large gap in the German defences between Caumont and Villers-Bocage. It was just asking for the British armour to rip through. The next day 7th Armoured was ordered to swing westwards to launch an attack on the enemy's left and exploit the gap. The forward tanks of 7th Armoured moved to within six miles of the hilltop town of Villers-Bocage before stopping for the night.

The next morning – 13 June – the leading tanks of 7th Armoured entered Villers-Bocage, to be greeted with delight by the French citizens. So unexpected was the British arrival that German billeting officers were still visiting houses to find accommodation for German soldiers in the town. Suddenly there was an air of festivity and some of the

British crews dismounted as if confident there were no Germans in the vicinity. They began chatting to the local people and accepting flowers and cakes. Montgomery, delighted by news of 7th Armoured's progress, signalled their success to England. But everyone was being lulled into a false sense of security. Unknown to them a single Tiger tank commanded by the leading tank 'ace' of the Second World War, Captain Michael Wittman, was approaching. Wittman stopped his tank and surveyed the British columns ahead of him. It seemed like peacetime manoeuvres, with British tank crews brewing up and enjoying the pleasant sunny weather. He was amazed at how complacent they all were. 'They're acting as if they've won the war already,' said Wittman's gunner. Wittman looked down and replied, 'We're going to prove them wrong.'

Suddenly Wittman's Tiger roared towards the stationary British Cromwell tanks. One after another Wittman pumped shells into each of the British tanks, leaving them flaming wrecks. Bursting through one remaining Cromwell, Wittman emerged in the main street of Villers-Bocage like one of the horsemen of the Apocalypse. He next met three more tanks of the County of London Yeomanry and destroyed each of them, then missed another Cromwell which disappeared into a garden. A British Sherman hit Wittman's Tiger several times without penetrating its armour. Wittman replied by blowing apart a building behind which the Sherman was sheltering, deluging the Sherman with its debris. Wittman then destroyed one more Cromwell and reversed away to reload and refuel.

Meanwhile, the rest of 'A' squadron, commanded by Lieutenant-Colonel Lord Cranley, was being attacked by four other Tigers from Wittman's command. Within a short space of time Cranley's tanks had been destroyed and himself taken prisoner. Wittman, having crushed the vanguard of 7th Armoured, now joined other Tiger tanks from the 2nd Panzer Division in their attacks on the British forces around Villers-Bocage. However, Wittman's luck had run out and the British ambushed him, destroying his Tiger, though he made his escape unhurt. But with more German troops moving into the area, 7th Armoured were forced to pull back what remained of their spearhead to the town of Tracy-Bocage, two miles away to the west. In one morning – and mainly at the hands of Wittman himself – they had lost 25 tanks and 28 armoured vehicles. It had been a shocking defeat. German general Fritz Kraemer, with due understatement, acknowledged Wittman's unprecedented achievement: 'The idea of placing in readiness the five Tiger tanks which had been left intact during the enemy air attacks had produced good results.' An RAF bombing attack had to be called in to flatten Villers-Bocage. On the following day 7th Armoured's progress was again held up by the devastating fire of the hand-held *Panzerfaust* anti-tank gun. Without adequate infantry support – inexplicably Bucknall had failed to ask 2nd Army for reinforcements – the British tanks were practically helpless against this weapon.

The British had failed to take the opportunity they had been offered and had allowed the Germans to close the gap in their lines that 7th Armoured had been aiming to exploit. British tank tactics had been amateurish throughout, with forward units travelling far ahead of any infantry support. The result was that whole columns of tanks could be held up by small groups of German infantry with anti-tank guns. The tanks' high explosive shells were quite inappropriate for use against infantry either well dug-in or highly mobile.

During the fighting inside Villers-Bocage, 7th Armoured had let Wittman run rings round them, destroying tanks at will. There were now serious questions being asked about the division's fighting spirit. There was a suspicion that many of the tank crews felt that they had done their fighting in the desert against Rommel and that it was somebody else's turn now. Fundamentally such attitudes only flourish under slack and uninspiring leadership. Risk avoidance was the name of the game and battle shyness was strongly built into these veteran units. Part of the trouble was the fact that tank crews had little confidence in their equipment. They felt that the Cromwell and Sherman tanks that they were using were in no way a match for the German Tigers. But there were ways of dealing with Tiger tanks and the feeble display of the tank crews facing Wittman had revealed a lack of talented leadership. The Germans had it and the British did not. As far as Montgomery was concerned the fault rested with the divisional and corps commanders, Erskine and Bucknall, and both men were subsequently sacked. As General Dempsey later wrote, '7th Armoured Division was living on its reputation and the whole handling of the battle was disgraceful,' Wittman had punished the British for their complacency. No war is over until the last bullet is fired. As far as Wittman and men like him were concerned, that last bullet would be fired by Germany. It is interesting to note that before Wittman died, killed near Falaise in early August 1944, he was credited with a personal score of 138 tank and self-propelled gun 'kills' as well as 132 anti-tank guns.

THE BATTLE OF HURTGEN FOREST (1944–5)

For six months during the winter of 1944–5 eight American infantry and two armoured divisions fought one of the bloodiest and most pointless battles of the Second World War. Its sheer futility caused one general to call it 'our Passchendaele', while to the ordinary GI it was simply known as 'the death factory'. The regiments that went into the 'green hell' of Hurtgen Forest lost at least 50 per cent of their men, and in two cases 100 per cent. Hurtgen just swallowed them up and spat out the bones.

The idea of capturing the Hurtgen Forest had been taken too lightly. It was of no strategic value. Although it was on the flank of General Collins' VIII Corps as it pressed on towards the German frontier, the forest could easily have been sealed off. Nobody needed to go in there. But once they did, and were well and truly defeated, the American top brass began to see Hurtgen in a different light. There may not be a strategic purpose to taking the forest but no American commander could allow his boys to be defeated and let the enemy get away with it. There was the prestige of the American army to be considered. The forest would have to be taken, whatever the cost. This was military madness. By the time the Americans came to realize it they had fought the longest battle in American history, suffered enormous losses and let down their British and Canadian allies so badly that it cost them 16,000 unnecessary casualties. The Germans never understood the attraction of Hurtgen for the Americans. In the forest fighting the Americans surrendered all the tactical advantages of air power. American General James Gavin later wrote, 'For us the Hurtgen was one of the most costly, most unproductive, and most ill-advised battles that our army has ever fought.'

The danger within the Hurtgen Forest was that it had been fortified by the Germans until it resembled a dragon's lair – it was bristling with mines, described as 'dragon's teeth'. There were concrete pillboxes, bunkers and wire defences. It was a place to go round, go over, leave alone – but not go through. It was a killing ground *par excellence*. Low-quality German troops in the Hurtgen could frustrate the plans of a Napoleon. But in September 1944 General Joseph Collins, 'Lightning Joe', decided to send an American division into the Hurtgen to root out the defenders. It is doubtful if Collins had really thought carefully about what he was doing. Why otherwise would he surrender all the advantages of mobility and send his men to fight tree by tree as if they were hunting Indians? Charles Whiting has called his decision perhaps the greatest mistake made by the Americans during the entire campaign from Normandy to Berlin. After Collins had made the first mistake all the divisional generals jumped on the bandwagon, in some kind of macho belief that their men could succeed where everyone else had failed. Divisions entered, lasted two weeks, and came out broken, with half their men killed or wounded and most of the rest ruined by combat fatigue. The shades of countless First World War generals must have smiled approvingly. And yet it was not even good attrition, for in return the Germans suffered scarcely any casualties at all. Their soldiers in the Hurtgen were second-class or even lower in quality. Many were old men, some were 'stomach cases' who needed special diets, others had glass eyes or wooden legs, yet others were mere boys. But these were good enough to press the trigger of a machine gun and mow down the Americans as they blundered about in the minefields. At first the Americans had smirked at the thought that Hitler was having to scrape the barrel. But the smile soon died in the Hurtgen. As one GI said, 'I don't care if the guy behind the gun is a syphilitic prick who's a hundred years old – he's still sitting behind eight feet of concrete and he's still got enough fingers to press triggers and shoot bullets.'

First, one of America's best infantry divisions went in – the 9th – 'proud and arrogant young men', who had always succeeded before in every task they had been set – until the Hurtgen. Nobody had bothered to train the 9th for forest warfare, where they would soon be beyond central control, where tanks could not support them and where the enemy might be all round them at any moment. Even the

American soldiers captured by German paratroops during the German Ardennes offensive, December 1944. For those American commanders who were confident of a speedy advance into Germany, the reverses they suffered during the battle of the 'Bulge' came as an unpleasant surprise.

mine detectors they had could not detect the German 'Schu' mines. On 6 October, the men from the 9th disappeared into the dark and dense forest only to find that the Germans were waiting for them. Every bunker and pillbox was occupied and every gun fired as the Americans came within range. The 9th were massacred.

The American generals soon came to the conclusion that if they were going to get bogged down in the Hurtgen Forest they had better find a reason for it. It was insufficient to pretend they were guarding Collins' flank and it would not go down well back home in America if it was admitted that they were fighting there just for prestige and to save a few reputations. And so a new reason was found. The American target now became the dams on the River Roer. If these dams were opened by the Germans much of the American advance could be stopped by floodwater. Now there was a reason for fighting in the Hurtgen Forest. However, none of the soldiers in the forest were told about the dams. They were there to fight Germans, and that was enough. Apparently the dams were never even mentioned at HQ briefings until November 1944. They were an afterthought. But they were a good enough excuse for what was otherwise a pointless operation.

On 24 October what was left of the 9th Division emerged from the forest. The survivors looked as if they had been to hell and back and had posed for Hieronymous Bosch on the return journey. The division's 60th Infantry Regiment had been wiped out almost to a man. Their replacements – 28th Division – now came in to bat, brimming with confidence. Soon their officers were falling like horse chestnuts in autumn. First the divisional commander was sacked. His replacement was shot through the head after two hours in charge. The next to take his turn was the tough – and controversial – Norman Cota. Cota intended to get things done and when he was told to take the village of Schmidt in the middle of the forest – which the 9th had not been able to do – he said it was as good as done. But talk was easy and Cota's boasting was going to cost the lives of thousands of his men. Yet none of their names would become as famous in American history as that of one who survived, Eddie Slovik. Slovik was soon going to achieve a unique distinction: the only man shot for desertion by the US army in the Second World War.

Cota's division made a bad start in Hurtgen. The 'American Luftwaffe' did what it could to help by carrying out a bombing of the forest. All they

Margaret Thatcher pays her last respects at the grave of Lieutenant-Colonel 'H' Jones, who died during the battle of Goose Green, posthumously winning the Victoria Cross.

Chiefs of Staff, along with Admiral Fieldhouse, were insistent that a successful land engagement was a valuable means of establishing a psychological superiority over the Argentinians. They believed that the enemy was weak and would present only token resistance. The politicians agreed, stressing that once the press turned against the war everyone might as well pack up and come home. But supposing they were wrong and the battle was lost? Did the Chiefs of Staff and the War Cabinet have any conception of the real situation at San Carlos? Under constant air raids as they were, the men at the front were not as convinced as the backseat drivers in London that the Argentinians were as weak as was popularly supposed.

On 26 May Thompson received an unequivocal order to attack Goose Green – or else. High political stakes rode on the success of this mission, which makes it all the more surprising that it was entered into with such scant superiority in men and equipment. If the battle had been lost the consequences for the overall campaign could have been disastrous.

Available intelligence reports suggested that Goose Green was held by just one weak Argentinian battalion or about 400 men, and that 2 Para, consisting of 450 men under Colonel 'H' Jones, with the support of three 105-mm guns, and the naval gunnery of HMS *Arrow*, would be strong enough to take the settlement. However, British intelligence had underestimated the Argentine garrison, which, in fact, consisted of over 600 fighting men as well as a substantial number of non-combatants, including air force personnel. Denied Scorpion and Scimitar light tanks, as petrol was short and the terrain might prove too difficult for them, the British were opting for a straightforward infantry action, in which they relied on their superior tactics and leadership to offset the Argentinian advantage in numbers and position. Surprisingly, the opportunity was not taken to reinforce 2 Para, an unnecessary risk in view of the importance being placed on the battle as a 'make or break' encounter.

General Menendez, the overall Argentinian commander, had brought in reinforcements to Goose Green once it was obvious that the British intended to attack it. Although the Argentinians had heard the BBC report that morning announcing that 2 Para was within five miles of Darwin they regarded it as a bluff. Nobody could make a mistake like that in wartime – it had to be intentional. So the final Argentinian garrison, commanded by Lieutenant-Colonel I. Piaggi, was slightly increased from 500 to 650 fighting men, of whom some were good-quality troops, well dug-in and far better than was being allowed for in London. The British troops were now heading for an encounter with an enemy that outnumbered them by three to two, a far cry from textbook requirements for an attacking force. But the British had no intention of doing things by the book. The capture of an Argentine Land-Rover with three occupants soon alerted 2 Para to the unpalatable fact that they were facing more than just the expected weak battalion.

In fact, the Argentinian garrison fought with determination throughout the battle, belying a description of them as 'demoralized and unmotivated'. They poured a heavy artillery and mortar fire on the paratroopers as they moved down the isthmus. Nevertheless, the sheer professionalism of 2 Para eventually won through though at a tragic cost. Their Colonel, 'H' Jones – later to be awarded a posthumous VC – was killed in the battle for Darwin Hill and in all 17 men died and 35 were wounded. The Argentinians lost 250 dead and missing and about 150 wounded. The most surprising fact from 2 Para's point of view was the size of their haul of prisoners: 1,007 – over twice their own total strength and far more than they had been led to believe were even in the area.

The intelligence failure which resulted in 2 Para being sent into action against such odds was a serious one. The outcome of the battle could have been disastrous and would have crowned a confused operation fought for political rather than strategic reasons. With hindsight it seems incredible that Britain was willing to fight her first battle on the Falklands with such a small infantry force. To rely on such a small attacking force was to run the risk of a military setback that could have sent morale plummeting in Britain. A British defeat might even have intensified moves to force both sides to agree to a ceasefire.

The attack on Goose Green reflected haste and underestimation of the enemy by those who set it in motion, redeemed only by the brilliant performance of 2 Para. It was an example of the sort of 'backseat-driving' that had led to disasters during the Second World War. The politicians and service-chiefs, deeply alarmed by the losses in San Carlos, wanted immediate action from the army to ease their political burden. The result was that a battle was fought not for military reasons but for political and propaganda purposes. The dangers of such an action were obvious. Defeat at Goose Green would have assumed a more than military importance. It would have called into question the whole strategy of sending an under-equipped expedition to the tip of South America in pursuit of aims that a post-colonial Britain should have outgrown. In the end the politicians relied on the heroism and superb fighting qualities of British troops to see them through. To demand an operation like Goose Green in the face of opposition from the man on the spot was to abuse political responsibility.

INDEX